Wissenschaftliche Reihe Fahrzeugtechnik Universität Stuttgart

Reihe herausgegeben von
M. Bargende, Stuttgart, Deutschland
H.-C. Reuss, Stuttgart, Deutschland
J. Wiedemann, Stuttgart, Deutschland

Das Institut für Verbrennungsmotoren und Kraftfahrwesen (IVK) an der Universität Stuttgart erforscht, entwickelt, appliziert und erprobt, in enger Zusammenarbeit mit der Industrie, Elemente bzw. Technologien aus dem Bereich moderner Fahrzeugkonzepte. Das Institut gliedert sich in die drei Bereiche Kraftfahrwesen, Fahrzeugantriebe und Kraftfahrzeug-Mechatronik. Aufgabe dieser Bereiche ist die Ausarbeitung des Themengebietes im Prüfstandsbetrieb, in Theorie und Simulation. Schwerpunkte des Kraftfahrwesens sind hierbei die Aerodynamik, Akustik (NVH), Fahrdynamik und Fahrermodellierung, Leichtbau, Sicherheit, Kraftübertragung sowie Energie und Thermomanagement – auch in Verbindung mit hybriden und batterieelektrischen Fahrzeugkonzepten.

Der Bereich Fahrzeugantriebe widmet sich den Themen Brennverfahrensentwicklung einschließlich Regelungs- und Steuerungskonzeptionen bei zugleich minimierten Emissionen, komplexe Abgasnachbehandlung, Aufladesysteme und -strategien, Hybridsysteme und Betriebsstrategien sowie mechanisch-akustischen Fragestellungen.

Themen der Kraftfahrzeug-Mechatronik sind die Antriebsstrangregelung/Hybride, Elektromobilität, Bordnetz und Energiemanagement, Funktions- und Softwareentwicklung sowie Test und Diagnose.

Die Erfüllung dieser Aufgaben wird prüfstandsseitig neben vielem anderen unterstützt durch 19 Motorenprüfstände, zwei Rollenprüfstände, einen 1:1-Fahrsimulator, einen Antriebsstrangprüfstand, einen Thermowindkanal sowie einen 1:1-Aeroakustikwindkanal.

Die wissenschaftliche Reihe „Fahrzeugtechnik Universität Stuttgart" präsentiert über die am Institut entstandenen Promotionen die hervorragenden Arbeitsergebnisse der Forschungstätigkeiten am IVK.

Reihe herausgegeben von

Prof. Dr.-Ing. Michael Bargende
Lehrstuhl Fahrzeugantriebe
Institut für Verbrennungsmotoren und
Kraftfahrwesen, Universität Stuttgart
Stuttgart, Deutschland

Prof. Dr.-Ing. Jochen Wiedemann
Lehrstuhl Kraftfahrwesen
Institut für Verbrennungsmotoren und
Kraftfahrwesen, Universität Stuttgart
Stuttgart, Deutschland

Prof. Dr.-Ing. Hans-Christian Reuss
Lehrstuhl Kraftfahrzeugmechatronik
Institut für Verbrennungsmotoren und
Kraftfahrwesen, Universität Stuttgart
Stuttgart, Deutschland

Weitere Bände in der Reihe http://www.springer.com/series/13535

Marlene Wentsch

Analysis of Injection Processes in an Innovative 3D-CFD Tool for the Simulation of Internal Combustion Engines

 Springer Vieweg

Marlene Wentsch
Stuttgart, Germany

Dissertation, University of Stuttgart, 2018

D93

Wissenschaftliche Reihe Fahrzeugtechnik Universität Stuttgart
ISBN 978-3-658-22166-9   ISBN 978-3-658-22167-6   (eBook)
https://doi.org/10.1007/978-3-658-22167-6

Library of Congress Control Number: 2018943323

Springer Vieweg

This Springer Vieweg imprint is published by the registered company Springer Fachmedien Wiesbaden
GmbH part of Springer Nature
The registered company address is: Abraham-Lincoln-Str. 46, 65189 Wiesbaden, Germany

Für Mama - Danke!

Preface

This work was realized during my activities as a research associate at the Institute of Internal Combustion Engines and Automotive Engineering (IVK) of the University of Stuttgart under the supervision of Prof. Dr.-Ing. Michael Bargende.

My deep gratitude goes to Prof. Dr.-Ing. Michael Bargende for providing the opportunity for this work. His scientific supervision and extensive support were particularly valuable for me. Working with him helped me identifying my strengths and contributed positively to my personal and professional development.

I also want to thank Prof. Dr.-Ing. Giorgio Rizzoni from Ohio State University (USA) for the co-referent of my work as well as many informative and inspiring discussions, also providing insights into the impressive work of the Center for Automotive Research (CAR).

My work was supported by various collaborations in the past years. Therefore, my sincere appreciation for the pleasant and successful cooperations goes to Dr.-Ing. Donatus Wichelhaus (Volkswagen Motorsport GmbH), Christian Pötsch (Volkswagen AG) and Daniel Koch (TU München) as well as Dr.-Ing. David Lejsek, Dr.-Ing. Paul Jochmann, Dr.-Ing. Fabian Köpple and Dr.-Ing. Dimitri Seboldt (Robert Bosch GmbH).

I am very grateful for my team members Dr.-Ing. Marco Chiodi, Oliver Mack (aka Meshelangelo) and Andreas Kächele for the always professional and yet fun collaboration as well as constant (moral) support, especially in the last phase of this work. Furthermore, I would like to thank all colleagues at IVK/FKFS for inspiring scientific discussions and a wonderful working atmosphere.

My warmest gratitude goes to my beloved family Elke and Timea Wentsch, Florian "Fetti" Heilmeier and my wonderful friends - thank you for your support, encouragement and unconditional love.

Ludwigsburg   Dipl.-Math. Marlene Wentsch

Contents

Preface VII

Figures XIII

Tables XVII

Abbreviations XIX

Symbols XXI

Abstract XXV

Kurzfassung XXIX

1   **Introduction** 1
     1.1   Virtual Engine Development 1
     1.2   The Role of 3D-CFD Simulations 3
     1.3   Importance of Injection Simulation 4

2   **Simulation of Internal Combustion Engines** 7
     2.1   Real Working Process Calculation 7
     2.2   1D-CFD Simulation 9
     2.3   3D-CFD Simulation 11
         2.3.1   Fundamental Equations 11
         2.3.2   Turbulence Modeling 17
         2.3.3   Application Possibilities and Limits 18

3   **The 3D-CFD Tool QuickSim** 21
     3.1   The Purpose of QuickSim 21
     3.2   Features of QuickSim 24
         3.2.1   Fast Analysis 3D-CFD Tool 24
         3.2.2   Full-Engine 3D-CFD Simulation 26
         3.2.3   Simulation of Successive Engine Operating Cycles 28
         3.2.4   Information Exchange Layout 30
     3.3   Development History and Portfolio of QuickSim 31
     3.4   Status Quo of Injection Modeling with QuickSim ... 32

3.5   Research Objective... 34

4   Fundamentals of Spray Modeling and Simulation......................... **35**

4.1   Fuel Properties and Specifications 35
        4.1.1   Liquid Fuel - Gasoline.. 35
        4.1.2   Gaseous Fuel - Compressed Natural Gas 36
4.2   Numerical Injection Definition.. 37
        4.2.1   Liquid Fuel Injection... 39
        4.2.2   Gaseous Fuel Injection 42
4.3   Spray Breakup Mechanisms 44
4.4   Droplet Evaporation 49

5   Utilized Engine Models **51**

5.1   Injection Chamber... 51
5.2   Turbocharged 4-Cylinder DISI-Engine 53
5.3   Single-Cylinder DISI-Engine.. 55
5.4   Turbocharged 2-Cylinder DI Weber MPE 850 DOHC 58

6   Numerical Boundary Conditions **61**

6.1   Numerical Stability and Convergence 65
        6.1.1   Injector Position 65
        6.1.2   Droplet Initialization Domain..................................... 66
        6.1.3   Lagrangian Droplet Approach 68
6.2   Spatial Discretization... 71
        6.2.1   Liquid Fuel 72
        6.2.2   Gaseous Fuel 79
6.3   Temporal Discretization 82
        6.3.1   Liquid Fuel 83
        6.3.2   Gaseous Fuel 89

7   Liquid Fuel Modeling **91**

7.1   Single-Component Fuel Modeling 91
        7.1.1   Literature Approaches 91
        7.1.2   Synthetic Gasoline Model 92

7.1.3   Limitations in the Application of Single-Component
Fuel Models 95
7.2   Multi-Component Fuel Modeling 97
7.2.1   Implementation of Multi-Component Fuel Models......... 97
7.2.2   Literature Approaches for Gasoline 100
7.2.3   QuickSim Specific Multi-Component Modeling 101
7.3   Application Example: Spray Targeting and Evaporation 105

8   Parametrization of Injector Properties...................................**119**

8.1   Injector Geometry and Manufacturing 119
8.2   Injector Mounting and Opening..128

9   Conclusion and Outlook...**141**

Bibliography 143
Appendix...151

Figures

1.1   Influencing parameters on the mixture formation process.................4

1.2   Main influencing parameters of interest 6

2.1   CPU time comparison of different calculation/ simulation tools 7

2.2   Combustion process modeling 9

2.3   1D-CFD discretization and calculation scheme............................ 10

2.4   Central computation point P 16

2.5   Modeling approaches for the description of a turbulent flow field 18

2.6   Comparison of spray propagation using a regular mesh structure and an adaptively refined mesh with injector nozzle resolution......... 20

3.1   CPU-time in dependence of the cell discretization length............... 25

3.2   CPU-time classification of QuickSim 25

3.3   Definition of boundary conditions at intake and exhaust mesh endings.. 27

3.4   3D-CFD simulation domain extension 27

3.5   Comparison of multi-cycle simulations for single-cylinder and full engine domain.. 29

3.6   Information exchange layout of QuickSim.................................. 30

4.1   3D-CFD mesh including spark plug and injector tip geometry 38

4.2   Spray angle defintion 38

4.3   Definition of L_{min} and L_{max} 40

4.4   Liquid fuel injection modeling... 42

4.5   Gaseous fuel injection modeling... 43

4.6   Spray breakup regimes.. 45

4.7   Spray atomization mechanisms 46

4.8   Secondary breakup mechanisms 48

5.1   3D-CFD mesh of the injection chamber 52

5.2   3D-CFD mesh of a turbocharged 4-cylinder DISI-engine............... 54

5.3   3D-CFD mesh of a single-cylinder DISI-engine........................... 56

5.4   Extended 3D-CFD mesh of a single-cylinder DISI-engine 57

5.5   Extended 3D-CFD mesh of a 2-cylinder DISI-engine.................... 59

6.1   Macroscopic fuel spray properties 61

6.2   Calibration result of gasoline injection, hollowcone injector 62

6.3   Calibration result of methane injection 63

6.4   Dependence of spray propagation on injector position.................... 65

6.5   Variation of the droplet initialization domain, 0.1 ms a. SOI............ 67

6.6   Variation of the droplet initialization domain, 0.5 ms a. SOI............ 67

6.7   Variation of the droplet initialization domain, 1.0 ms a. SOI............ 68

6.8   Influence of droplets per parcel definition on spray propagation 69

6.9   Influence of droplets per parcel definition on spray evaporation........ 70

6.10 Influence of spatial cell discretization on liquid fuel propagation,
      T_{ch}=293 K 72

6.11 Influence of spatial cell discretization on evaporated fuel
      propagation, T_{ch}=293 K... 73

6.12 Influence of spatial cell discretization on liquid fuel propagation,
      section plot, T_{ch}=293 K 74

6.13 Influence of spatial cell discretization on liquid fuel propagation,
      T_{ch}=400 K 75

6.14 Influence of spatial cell discretization on liquid fuel propagation,
      section plot, T_{ch}=400K 75

6.15 Influence of spatial cell discretization on evaporated fuel
      propagation, T_{ch}=400 K... 76

6.16 Influence of spatial cell discretization on liquid fuel propagation,
      5-hole injector... 77

6.17 Influence of spatial cell discretization on liquid fuel evaporation,
      5-hole injector... 78

6.18 Influence of spatial cell discretization on gaseous fuel propagation ... 80

6.19 Schematic illustration of the spatial discretization influence on
      gaseous fuel initialization and propagation 81

6.20 Influence of temporal discretization on liquid fuel propagation,
      $\Delta t \leq 1/80$ ms 83

6.21 Influence of temporal discretization on liquid fuel propagation 84

6.22 Influence of temporal discretization on liquid fuel evaporation.......... 86

6.23 Influence of temporal discretization on liquid fuel evaporation,
      engine simulation... 87

6.24 In-cylinder lambda distribution at IP for different time increments.... 88

6.25 Influence of temporal discretization on gaseous fuel propagation...... 89

7.1   Approximation of fuel density with n-undecane............................. 93

7.2   Approximation of fuel viscosity with cyclopentane 94

7.3   Polynomial saturation pressure approximation............................. 95

7.4   Boiling curve reference racing gasoline 96

7.5   Fuel decomposition due to varying component volatility 96

7.6   Comparison of fuel modeling approaches.................................... 98

7.7   Simulated evaporation behavior of a 10-component fuel model.......100

7.8   Comparison of experimental saturation pressure curve with
      multi-component model of racing gasoline 102

7.9   QuickSim multi-component droplet initialization........................104

7.10 Diesel and gasoline fuel spray propagation 106

7.11 Calibration result of gasoline injection, 5-hole injector.................107

7.12 Saturation pressure comparison single-component fuel models 108

7.13 Saturation pressure comparison multi-component fuel models 108

7.14 Influence of single-component fuel models on spray propagation
      and evaporation, STP...109

7.15 Influence of multi-component fuel models on spray propagation
      and evaporation, STP...110

7.16 Influence of single-component fuel models on spray propagation
      and evaporation, T_{ch}=373 K..112

7.17 Influence of multi-component fuel models on spray propagation
      and evaporation, T_{ch}=373 K..113

7.18 Influence of single-component fuel models on spray propagation
      and evaporation, p_{ch}=2.5 bar 115

7.19 Influence of multi-component fuel models on spray propagation
      and evaporation, p_{ch}=2.5 bar 116

8.1   Geometrical deviations of different injector variants, frontal view....120

8.2   Geometrical deviations of different injector variants, side view 121

8.3   Comparison of test bench images to simulated spray with
      geometrical injector definition 123

8.4   Experimentally and numerically derived spray jet axes in
      comparison to the geometrical injector targeting 124

8.5   Comparison of test bench images to simulated spray with modified
      and geometrical injector definition, frontal view 125

8.6   Comparison of test bench images to simulated spray with modified
      and geometrical injector definition, side view..............................126

8.7   Comparison of the fuel distribution at IP employing different
       numerical injector geometries 127
8.8   Comparison of the tumble level employing different numerical
       injector geometries 128
8.9   HC measurements in the exhaust gas and intake manifold 130
8.10 Schlieren images of different injector mounting positions 131
8.11 3D-CFD mesh modification for different injector mounting
       positions 131
8.12 Comparison of Schlieren images to simulated spray propagation
       for mounting depth of -2 mm (constant α) 132
8.13 Comparison of Schlieren images to simulated spray propagation
       for mounting depth of -2 mm (variable α) 133
8.14 Comparison of Schlieren images to simulated spray propagation
       for mounting depth of ± 0 mm 135
8.15 Comparison of Schlieren images to simulated spray propagation
       for mounting depth of -4 mm 136
8.16 Simulation results of different injector mounting positions 137
8.17 Comparison of fuel distribution for different injector mounting
       positions 138
8.18 Visualization of fuel scavenging for different injector mounting
       positions 139

Tables

4.1 Examples for injector type specific geometry parameters 39

5.1 Standard simulation configuration of the injection chamber 52

5.2 WRC engine specifications 53

5.3 Standard simulation configuration of the WRC engine 54

5.4 Single-cylinder research engine specifications 55

5.5 Standard simulation configuration of the single-cylinder engine 58

5.6 Weber engine specifications 59

5.7 Standard simulation configuration of the Weber engine 60

6.1 CPU-time comparison for different droplet per parcel definitions 70

6.2 CPU-time comparison for different degrees of calculation mesh refinement 79

6.3 Run time comparison for different degrees of calculation mesh refinement 79

6.4 Engine cycle comparison for different time steps Δt 88

7.1 Thermo-physical properties of racing gasoline (RON = 100.3), n-heptane and iso-octane 92

7.2 5-component fuel model for a racing gasoline 103

A1.1 Implemented model environment 151

A3.1 Thermo-physical properties of 1c gasoline models 153

A3.2 Thermo-physical properties of 3c gasoline models 153

A3.3 Thermo-physical properties of 5c gasoline model 154

A3.4 Thermo-physical properties of 7c gasoline model 155

Abbreviations

0D	Zero-dimensional
1D	One-dimensional
3D	Three-dimensional
BDC	Bottom dead center
CA	Crank angle
CAD	Computer-aided design
CFD	Computational fluid dynamics
CNG	Compressed natural gas
CPU	Central processing unit
DI	Direct injection
DNS	Direct numerical simulation
E	Exhaust
EGR	Exhaust gas recirculation
FIA	Fédération Internationale de l'Automobile
FKFS	Forschungsinstitut für Kraftfahrzeuge und Fahrzeugmotoren Stuttgart
FTDC	Firing top dead center
HCCI	Homogeneous charge compression ignition
I	Intake
ICE	Internal combustion engine
IP	Ignition point
IVC	Intake valve closing
IVK	Institut für Verbrennungsmotoren und Kraftfahrwesen

LES	Large eddy simulation
LVK	Lehrstuhl für Verbrennungskraftmaschinen
MPI	Multi-point injection
PDA	Phase doppler analyzer
PFI	Port fuel injection
PRF	Primary reference fuel
RANS	Reynolds average navier stokes
RDE	Real driving emissions
RON	Research octane number
rpm	Revolutions per minute
SI	Spark-ignition
SMD	Sauter mean diameter
SOI	Start of injection
SPI	Single-point injection
STP	Standard temperature and pressure
TDC	Top dead center
VOL	Valve overlap
WOT	Wide open throttle
WPC	Real working process calculation
WRC	World Rally Championship

Symbols

Latin Letters

b_i	Indicated specific fuel consumption	g/kWh
$BMEP$	Break mean effective pressure	bar
C	Courant number	-
c_f	Source term in the conservation equations	-
c_p	Specific heat capacity	J/kgK
d_D	Droplet diameter	m
$d_{D,s}$	Stable droplet diameter	m
e	Specific energy	J/kg
F	General extensive variable of conservation equations	-
f	General density variable of conservation equations	-
G	Potential gravitational energy	J/kg
$\vec{g}$	Gravity acceleration	m/s^2
h	Specific enthalpy	J/kg
H_E	Exhaust enthalpy	J
H_I	Intake enthalpy	J
h_v	Heat of vaporization	J/kg
$\bar{\bar{I}}$	Unit matrix	-
$IMEP$	Indicated mean effective pressure	bar
$\vec{J}_i$	Diffusion mass flux of species i	kg/m^{2}s
$\vec{J}_q$	Heat conduction	W/m^{2}s
l_c	Cell discretization length	mm
$L_{\max}$	Maximal initialization distance from injector coordiate system	mm
$L_{\min}$	Minimal initialization distance from injector coordinate system	mm
$\dot{m}$	Mass flow rate	mg/s
m	Mass	kg

m_A	Air mass	kg
m_B	Fuel mass	kg
m_b	Mass in the burned zone in the cylinder	kg
m_C	Cylinder mass	kg
m_E	Exhaust mass	kg
M_i	Molar mass of species i	kg/kmol
m_I	Intake mass	kg
m_i	Mass of species i	kg
m_L	Leakage mass	kg
m_u	Mass in the unburned zone in the cylinder	kg
$\vec{n}$	Normal vector	-
N_D	Number of droplets per parcel	-
$N_{D,i}$	Number of droplets per parcel for component i	-
Oh	Ohnesorge number	-
$\bar{\bar{P}}$	Stress tensor	N/m^2
P	Central computation point	-
p	Pressure	bar
P_a	Axial spray penetration	mm
p_c	Critical pressure	Pa
p_{ch}	Injection chamber pressure	bar
p_{exh}	Pressure at exhaust boundary	bar
p_{inj}	Injection pressure	bar
p_{int}	Pressure at intake boundary	bar
p_{max}	Maximum cylinder pressure	bar
P_r	Radial spray penetration	mm
p_s	Saturation pressure	Pa
Q_B	Fuel heat release energy	J
q_r	Specific energy term due to radiation or magnetic fields	J/kgs
Q_W	Wall heat transfer	J
$Q_{W,b}$	Wall heat transfer in the burned zone in the cylinder	J
$Q_{W,u}$	Wall heat transfer in the unburned zone in the cylinder	J
r	correlation coefficient	-
Re	Reynolds number	-

R_s	Specific gas constant	J/kgK
S	Computation cell surface	-
s_f	Source term in the conservation equations	-
S_j	Sub-surface of S	-
T	Temperature	K
t	Time	s
T_b	Temperature of the burned zone in the cylinder	K
T_b	Boiling temperature	K
T_c	Critical temperature	K
T_{ch}	Injection chamber temperature	K
T_{exh}	Temperature at exhaust boundary	K
T_f	Fuel temperature	K
T_{int}	Temperature at intake boundary	K
T_u	Temperature of the unburned zone in the cylinder	K
U	Internal energy	J
u	Specific internal energy	J/kg
V	Volume	m^3
v	Average flow velocity	m/s
v	Injection velocity	m/s
$\vec{v}$	Flow velocity	m/s
V_b	Volume of the burned zone in the cylinder	m^3
$\vec{V}_i$	Diffusion velocity of species i	m/s
$\vec{v}_i$	Velocity of species i	m/s
v_{rel}	Relative velocity between droplet and surrounding medium	m/s
V_u	Volume of the unburned zone in the cylinder	m^3
W	Work	J
We	Weber number	-
w_i	Mass fraction of species i	kg/kg
$\vec{x}$	Position vector	m
z	Injector middle axis	-

Greek Letters

α	Spray jet angle	°

ε	Spray cone angle	°
η	Dynamic viscosity	Ns/m^2
η	Efficiency	-
η_f	Droplet dynamic viscosity	Ns/m^2
η_i	Indicated efficiency	-
γ	Hollowcone spray angle	°
Ω	Fixed volume	-
$\Delta\Omega$	Surface of volume Ω	-
ω_i	Molar formation rate of species i	$kmol/m^3s$
φ	Crank Angle	°
φ	Spray azimuth angle	°
$\vec{\Phi}_f$	Flux of the general density variable f in the conservation equations	-
ρ	Density	kg/m^3
ρ_g	Density of the droplet surrounding gas	kg/m^3
ρ_i	Density of species i	kg/m^3
σ	Surface tension	N/m
σ_f	Droplet surface tension	N/m
Δt	Time step	°CA or ms
τ	Characteristic droplet breakup time	s
θ	Spray zenith angle	°

Abstract

Despite the increasingly politically driven electrification of the powertrain, the automotive industry is striving to ensure that the internal combustion engine still remains an legitimate alternative (for example in a hybrid vehicle) in the distant future. The demand for high efficiency and low emissions, while at least keeping the level of performance, is therefore greater than ever before. The socio-political pressure as well as steadily shorter product life cycles require not only an optimization of the engine itself but also an efficiency increase in the organization of the development process. Increasing virtualization of the engine development offers great opportunities with regard to internal engine optimization possibilities, considering a large variety of variants while reducing competitively relevant financial and temporal effort.

The 3D-CFD tool QuickSim was developed at FKFS and IVK of the University of Stuttgart in order to enable comprehensive studies and virtual development of internal combustion engines. By specific modeling of the combustion, heat transfer, etc., a coarsening of the mesh structure and thus significant acceleration of engine calculations could be realized. Full-engine simulations over several successive cycles enable a holistic analysis of the flow field and thus reliable and predictive statements on development-relevant questions.

Since a well targeted injection of the fuel in sufficient quantity and at the right time has a decisive influence on the subsequent processes of ignition, combustion and pollutant formation, it is an important part of (virtual) engine development. 3D-CFD simulations can be particularly useful for the geometric injector design as well as the development of injection strategies.

The aim of the present work was a critical analysis and assessment along with corresponding optimization and extension of the injection modeling in Quick-Sim. Due to the large number of influencing parameters and interactions, the fuel injection and therewith fuel propagation and distribution are part of the most complex processes in an internal combustion engine. For this reason, injection is usually the subject to highly detailed numerical investigations. In the context of virtual engine development, however, such a time-intensive calculation of a subprocess is neither desired nor purposeful. Plausibility provided, the presented investigations as well as the assessment of individual analysis results are therefore primarily oriented towards conformity with QuickSim as fast response 3D-CFD tool.

The injection simulation is essentially characterized by the choice of fuel model, the parameterization of injection conditions and injector geometry as well as the numerical framework conditions. The latter represented the foundation and thus the starting point of the investigations. By means of extensive parameter variations, the impact on macroscopic spray properties as well as the fuel evaporation was analyzed. The use of a virtual injection chamber made it possible to assess the spray behavior free from external influences (limiting combustion chamber walls, charge motion, etc.). For the validation of reference simulations, optical images of the liquid and gaseous fuel injection with a hollowcone injector were used. Numerous numerical influencing factors could be identified, but the changes in numerical boundary conditions had a particularly large effect on the simulation of gaseous fuel. This, in contrast to liquid fuel, was highly sensitive to a change in the time step as well as mesh discretization. Particularly the latter caused a considerable change in the radial and axial spray penetration, whereby a coarse cell structure was unable to reproduce the defined injector geometry. The cell edge length thus needs to be limited for gaseous fuel applications. Furthermore, it was shown that even a continuous mesh refinement does not necessarily lead to an improvement in the spray calculation, but instead to a noticeable increase of CPU-time due to the associated smaller time steps. It is rather useful to specify an application-specific choice of temporal and spatial discretization, on the basis of which a thorough calibration of further injection parameters can be finally carried out.

Fuel modeling is usually subject to severe simplifications in the simulation of internal combustion engines. However, since the fuel properties considerably influence the spray breakup, fuel evaporation and mixture formation, different model approaches were implemented in QuickSim. Based on a comprehensive fuel analysis, it was possible to create a single-component as well as a discrete multi-component model and to analyze these together with selected literature approaches (also single- and multi-component) regarding their influence on the fuel spray propagation. It has been shown that the choice of model fuel, regardless of prevailing pressure and temperature conditions as well as the injector geometry, has almost no influence on macroscopic spray properties. However, a comparison of the evaporation behavior showed significant model (composition) dependencies. This implies that a multi-component modeling of the fuel, due to the representation of a boiling curve, can be more suitable for the simulation of strongly fuel-dependent phenomena, for example soot formation.

Using the example of injector geometry, it became obvious that not all phenomena within the fuel injection can be directly transferred into calculation models. This is in part due to the fact that mandatory model parameters are usually not available (injector dimensions, flow coefficients, etc.). Moreover, considerable deviations of the fuel jet propagation can also be caused by effects which are not of thermodynamic origin. Instead of the meticulous attempt for generally valid modeling (which usually leads to a considerable increase of CPU-time), an individual injector calibration, in preparation for engine simulations, is preferable.

In summary, the present study confirms that the characteristic length employed in QuickSim is also adequate and useful for complex processes such as fuel injection - given a suitable choice of injection parameters and boundary conditions. These are strongly dependent on the application as well as underlying questions, i.e. a case-specific calibration of the injector and fuel spray is essential. Only a well balanced interaction between the conditioned 3D-CFD mesh, adapted calculation models as well as corresponding model parameters and boundary conditions, enables a reliable and predictive simulation of injection processes in the context of engine development tasks with QuickSim.

Kurzfassung

Trotz der zunehmend politisch getriebenen Elektrifizierung des Antriebsstrangs bemüht sich die Automobilindustrie darum, dass der Verbrennungsmotor auch in ferner Zukunft noch als legitime Alternative (bspw. im Hybridverbund) gilt. Der Anspruch an hohe Effizienz und niedrige Emissionen bei mindestens gleichbleibendem Leistungsniveau ist somit größer denn je. Der gesellschaftspolitische Druck sowie stetig kürzere Produktlebenszyklen erfordern nicht nur eine Optimierung des motorischen Antriebs selbst sondern auch eine Effizienzsteigerung in der Gestaltung des Entwicklungsprozesses. Die zunehmende Virtualisierung der Motorentwicklung bietet hierbei große Chancen hinsichtlich innermotorischer Optimierungsmöglichkeiten unter Berücksichtigung einer großen Variantenvielfalt bei zeitgleicher Reduktion des wettbewerbsrelevanten Zeit- und Kostenaufwands.

Mit dem 3D-CFD Programm QuickSim wurde am FKFS bzw. IVK der Universität Stuttgart ein Werkzeug entwickelt, welches speziell in der Entwicklung von Verbrennungsmotoren Anwendung findet. Durch eine spezifische Modellierung der Verbrennung, des Wärmeübergangs, etc. konnte eine Vergröberung der Rechennetzstruktur und damit eine deutliche Beschleunigung der Motorberechnung realisiert werden. Vollmotorsimulationen über mehrere Zyklen ermöglichen eine ganzheitliche Betrachtung des Strömungsfelds und damit verlässliche, prädiktive Aussagen zu entwicklungsrelevanten Fragestellungen.

Da die zielgerichtete Einbringung des Kraftstoffes in ausreichender Menge und zum richtigen Zeitpunkt einen entscheidenden Einfluss auf die darauffolgenden Prozesse der Zündung, Verbrennung sowie Schadstoffbildung hat, ist diese wichtiger Bestandteil der (virtuellen) Motorentwicklung. Vor allem bei der geometrischen Auslegung des Injektors sowie der Entwicklung von Einspritzstrategien können 3D-CFD Simulationen sinnvoll eingesetzt werden.

Ziel der vorliegenden Arbeit war eine kritische Analyse und Beurteilung sowie entsprechende Optimierung bzw. Erweiterung der Einspritzmodellierung in QuickSim. Durch die Vielzahl von Einflussparametern und Wechselwirkungen

zählt die Kraftstoffeinspritzung, und damit verbunden die Kraftstoffausbreit-
ung und -verteilung, zu den komplexesten Vorgängen im Verbrennungsmotor.
Aus diesem Grund ist sie meist Gegenstand höchst detaillierter numerischer
Untersuchungen. Im Rahmen der virtuellen Motorentwicklung ist eine solch
zeitintensive Berechnung eines Teilprozesses jedoch weder gewünscht noch
zielführend. Plausibilität vorausgesetzt, orientieren sich die dargelegten Untersuchungen sowie die Beurteilung einzelner Analyseergebnisse in dieser Arbeit
daher primär an der Konformität mit QuickSim als schnelles 3D-CFD Berech-
nungsprogramm.

Die Simulation der Einspritzung wird im Wesentlichen durch die Wahl des
Kraftstoffmodells, der Parametrisierung von Einspritzbedingungen und Injektorgeometrie sowie der numerischen Rahmenbedingungen charakterisiert. Letz-
tere bildeten das Fundament und damit Ausgangspunkt der Untersuchungen.
Mittels umfangreicher Parametervariationen wurde die Beeinflussung von mak-
roskopischen Sprayeigenschaften sowie der Kraftstoffverdampfung analysiert.
Der Einsatz einer virtuellen Einspritzkammer ermöglichte es hierbei, das Spray-
verhalten frei von äußeren Einflüssen (begrenzende Brennraumwände, Ladungs-
bewegung, etc.) zu beurteilen. Zur Validierung der Referenzsimulationen wurden optische Aufnahmen der flüssigen sowie gasförmigen Einspritzung mit
einer A-Düse verwendet. Es konnte eine Vielzahl numerischer Einflussfaktoren
identifiziert werden, besonders deutliche Auswirkungen hatte die Änderungen
von numerischen Randbedingungen allerdings bei der Simulation von gas-
förmigem Kraftstoff. Dieser reagierte im Gegensatz zu flüssigem Kraftstoff
höchst empfindlich auf eine Modifikation der Zeitschrittweite sowie Netzdiskret-
isierung. Insbesondere Letzteres bewirkte eine erhebliche Änderung der ra-
dialen und axialen Strahlpenetration, wobei zu grobe Zellen nicht in der Lage
waren die definierte Injektorgeometrie wiederzugeben. Der Zellkantenlänge
ist damit für die gasförmige Einblasung eine prinzipbedingte Obergrenze ge-
setzt. Weiter konnte gezeigt werden, dass auch eine stete Netzverfeinerung
nicht zwangsläufig eine Verbesserung der Sprayberechnung bewirkt, durch die
damit einhergehenden kleineren Zeitschrittweiten jedoch einen signifikanten
Anstieg der Rechenzeit. Vielmehr ist eine anwendungsspezifische Wahl der
zeitlichen und räumlichen Diskretisierung sinnvoll, auf deren Basis schließ-
lich eine sorgfältige Kalibrierung der weiteren Einspritzparameter erfolgt.

Die Kraftstoffmodellierung unterliegt bei der Simulation von Verbrennungs-motoren mitunter den stärksten Vereinfachungen. Da die Kraftstoffeigenschaf-ten den Verlauf des Strahlzerfalls, der Verdampfung und Gemischbildung je-doch maßgeblich beeinflussen, wurden in QuickSim unterschiedliche Modell-ansätze implementiert. Auf Basis einer umfangreichen Kraftstoffanalyse war es möglich ein einkomponentiges sowie diskretes mehrkomponentiges Modell zu erstellen und diese gemeinsam mit ausgewählten Literaturansätzen (eben-falls ein- und mehrkomponentig) hinsichtlich ihres Einflusses auf die Kraft-stoffstrahlausbreitung zu analysieren. Es konnte gezeigt werden, dass die Wahl des Ersatzkraftstoffes, unabhängig von herrschenden Druck- und Temperatur-bedingungen sowie der Injektorgeometrie, nahezu keinen Einfluss auf mak-roskopische Sprayeigenschaften hat. Der Vergleich des Verdampfungsverhal-tens wies dagegen deutliche Modellabhängigkeiten auf. Dies impliziert, dass eine mehrkomponentige Modellierung des Kraftstoffes, aufgrund der Möglich-keit einen differenzierten Siedeverlauf darzustellen, für die Simulation stark kraftstoffabhängiger Phänomene, beispielsweise der Rußbildung, besser geeig-net ist.

Am Beispiel der Injektorgeometrie wurde deutlich, dass nicht alle bei der Kraftstoffeinspritzung auftretenden Phänomene direkt in Berechnungsmodelle überführt werden können. Dies liegt einerseits darin begründet, dass die not-wendigen Modellparameter meist nicht verfügbar sind (Injektormaße, Durch-flusskoeffizienten, etc.), andererseits werden erhebliche Abweichungen der Strahlausbreitung auch durch Effekte, welche nicht thermodynamischen Ur-sprungs sind, bewirkt. Anstelle des akribischen Versuchs einer allgemein gülti-gen Modellerstellung (welche im ungünstigsten Fall zu einer erheblichen Ver-längerung der Rechenzeit führt) ist daher eine individuelle Injektorkalibrier-ung, in Vorbereitung auf motorische Simulationen, vorzuziehen.

Zusammenfassend konnte durch die vorliegende Arbeit bestätigt werden, dass die in QuickSim implementierte charakteristische Länge auch für komplexe Prozesse wie die Kraftstoffeinspritzung ausreichend und sinnvoll ist - eine geeignete Wahl von Einspritzparametern und Randbedingungen vorausgesetzt. Diese sind stark von der Anwendung sowie zugrundeliegenden Fragestellung abhängig, daher ist eine fallspezifische Kalibrierung des Injektors und Kraft-stoffsprays unerlässlich. Nur eine abgestimmte Interaktion zwischen konditioniertem 3D-CFD Netz, angepassten Berechnungsmodellen sowie entsprechenden Modellparametern und Randbedingungen ermöglicht eine verlässliche und prädiktive Simulation von Einspritzprozessen im Rahmen motorischer Entwicklungsaufgaben mit QuickSim.

1 Introduction

1.1 Virtual Engine Development

The development of internal combustion engines is a very demanding task, which requires a deep understanding of thermodynamic processes and their influencing factors. The complexity and extensiveness, which has to be mastered, is additionally enhanced by constantly increasing requirements in terms of emissions, engine efficiency, performance and drivability as well as the introduction of new technologies like water injection and homogeneous charge compression ignition (HCCI) or test methods like real driving emissions (RDE). Shortened product life cycles and thus significantly reduced development times [47] moreover require time-saving processes in the development of new as well as the optimization of existing powertrain concepts. Today's automotive industry therefore works with a variety of development tools, which highly differ concerning their field of application, level of detail, predictive capability, resource consumption, etc. [16].

Experimental analyses can be performed holistically under real conditions at test bench facilities like single-cylinder and full engine units, equipped with comprehensive measurement technology (incl. optical devices). These tests are complemented by theoretical analysis tools, which allow a backward projection of the combustion process. This offers the possibility of evaluating parameters, which cannot be measured directly on a test bench: indicated efficiency η_i, indicated mean effective pressure $IMEP$ or indicated specific fuel consumption b_i, etc. [5]. Additional labor equipment, including injection chambers, mainly utilized for spray investigations, rapid compression machines for auto ignition tests, blowing test stands, etc. enable particularly targeted investigations of engine sub-processes.

Analyses on experimental devices provide a huge variety of information on the engine behavior. However, a complex matrix of requirements, which includes innumerable component variants and parameterization options, can nowadays no longer be mastered solely with test bench analyses [25]. Supported by continuously growing computing power and further development in simulation software, numerical simulation tools, i.e. computational fluid dynamics (CFD) codes, enable the investigation of numerous variants in spite of limited time and financial capacities. The engine operating cycle and its parameters can be evaluated locally and temporally resolved, providing a prognosis of the combustion process and engine characteristics [5]. In addition, a fundamental understanding of effects and internal engine phenomena can be built up easier, since these can be visualized with CFD simulations instead of extensive measurement effort, e.g. installation of an optical access. [25, 41, 79]

Simulations and practical tests, which are characterized by varying strengths and weaknesses, are both of great use in the development of modern engine concepts. Due to this, the crucial challenge for a successful engine development process cannot be a substitution of one method by another, but to create symbiotic developmental structures which benefit from their individual advantages while compensating major drawbacks [25]. [20] sees great potential for the optimization of the development process by successively increasing the use of simulation tools. In this way, the more distinctive variety of variants (also including more operating points for the RDE testing method) can be mastered with simultaneous reduction of development effort for each variant. Simulative approaches are gaining relevance particularly in the preliminary design, variant selection and evaluation of different engine concepts during the early concept stage, even before the production of a first hardware prototype ("Frontloading"). Without major financial risk, also unusual or novel concepts, which might not be implemented on the test bench, will be taken into account. An extensive matrix of virtually tested engine variants thereby increases the confidence in engine design decisions [75]. Further (optimized) variants can then be derived from the initial design, also using simulation tools. In this way, the extent of necessary experimental investigations can be significantly reduced. These then predominantly serve to validate the calculations, verify simulated concept solutions and moreover enable a quantification of engine parameters, in particular emission values [20, 25].

1.2 The Role of 3D-CFD Simulations

Given limited engine dimensions and weight, engine internal measures offer great potential for the increase of engine performance and efficiency as well as the reduction of exhaust emissions. Three-dimensional (3D) CFD simulations provide the most detailed temporal and spatial resolution of flow influenced processes inside internal combustion engines [40]. This is particularly important when considering and evaluating highly three-dimensional processes like the gas exchange (influenced by channel and cylinder head geometry), fuel injection and distribution (spray propagation, charge homogenization, local enrichment, wall attachment, etc.), combustion (significantly influenced by the induced turbulence) and emission behavior as well as respective geometry dependencies (e.g. of the channels, piston and injector). Targeted 3D-CFD simulations can therefore extensively support the engine development process, not only by improving the understanding of processes occurring within the engine but also by predicting the influence of engine geometry and control strategies on the engine performance over a wide range of operating points. Major advantages over test bench investigations can be seen in a better reproducibility of analyses and their boundary conditions [20], a direct view inside the engine [16] and the delivery of results within a comparatively short period of time [75], not least owing to constantly increasing computing power.

Due to the high complexity of engine processes, 3D-CFD simulations yet represent the most challenging approach for the investigation of fluid dynamics problems. Technical, physical and numerical know-how and experience is mandatory to interpret the calculation results correctly and to derive appropriate measures [47]. Constant critical monitoring of the results' plausibility is of utmost importance here. Depending on the tool and the (user) defined initial and boundary conditions, the reliability and predictive ability can be limited and need to always be questioned by the development engineer [41].

1.3 Importance of Injection Simulation

In the development of injection concepts, in particular direct injection (DI), a major goal can be seen in the adjustment of the fuel jet, its resulting interaction with the in-cylinder flow and the derivation of a spray targeting, which ensures a targeted distribution of fuel inside the combustion chamber [25]. A predictive description of the mixture formation directly affects the ignition and combustion (which in turn has essential influence on the formation of emissions and the engine performance) and is consequently necessary for a thorough virtual engine development [77, 78].

[44] provides a detailed overview of influencing factors on the mixture formation process. According to Figure 1.1, a numerical modeling (incl. boundary conditions) of the fuel injection process needs to consider not only the (engine) operating conditions but also fuel and injector specific characteristics as well as the in-cylinder charge motion.

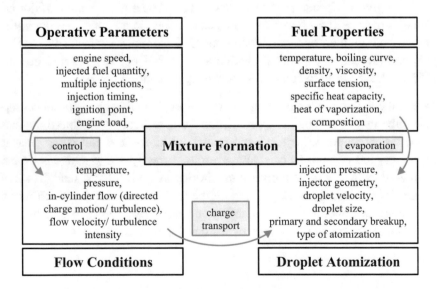

Figure 1.1: Influencing parameters on the mixture formation process [44]

Using the 3D-CFD tool QuickSim, which will be extensively discussed in Chapter 3, the engine operating conditions can be implemented without further ado. Moreover, a reliable calculation of the flow field, particularly inside the combustion chamber, can be ensured. Therefore, the focus of the presented work is put on the numerical description of fuel and injector specific parameters in order to enable the simulations of:

- Injector position:
  direct injection, port fuel injection (single-point or multi-point)
- Injector geometry:
  inward-opening injector (multi-hole or single-hole), outward-opening injector (hollowcone injector)
- Injection conditions:
  pressure, temperature, etc.
- Injection timing:
  number of injections, injection duration, end of injection
- Fuel type:
  gasoline, diesel fuel, ethanol, bio fuel, CNG, LPG, fuel blends

Since the simulation of the injector internal flow as well as the primary spray breakup are omitted in QuickSim, some of these properties need to either be empirically derived or modeled. Auxiliary variables are required as input or initial values for respective calculation models. These partly depend on one another or rather influence each other. The injection modeling is therefore based on an interdependency between calculation models and the definition of correct initial and boundary conditions.

The present work comprises a critical analysis and evaluation of the latter. Conformity with the 3D-CFD tool QuickSim is thereby taken into account as a decisive criterion. This entails not only the consideration of implemented calculation models but in particular of the main characteristic of QuickSim, being a fast response 3D-CFD engine development tool. The main task therefore encompasses the identification of a necessary, application dependent degree of detailing in the injection modeling and boundary conditions definition with regard to the simulation results quality, i.e. the predictability and reliability of the engine operating cycle calculation in QuickSim. For this purpose, the analysis is divided into three main areas of interest, illustrated in Figure 1.2: fuel modeling, injector and injection strategy implementation as well as numerical framework conditions.

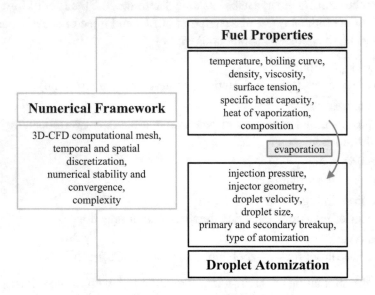

Figure 1.2: Main influencing parameters of interest

2 Simulation of Internal Combustion Engines

The virtual engine development is based on a well-considered integration of three main simulation tools: real working process calculation, one-dimensional (1D) CFD simulation and 3D-CFD simulation. Due to their strongly deviating temporal and spatial resolution, they have different modeling demands, input requirements and therefore fields of application. They highly differ in their degrees of complexity, predictive capability and computing time as illustrated in Figure 2.1. The specifics of each tool and their contribution to the development of internal combustion engines are in summary described below. More information can be taken from [16, 40, 72].

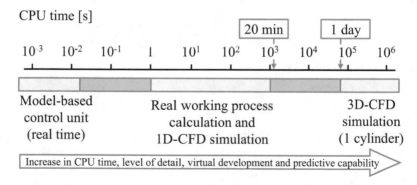

Figure 2.1: CPU time comparison of different calculation/ simulation tools [7]

2.1 Real Working Process Calculation

The analysis and calculation of in-cylinder thermodynamic processes is done by means of zero-dimensional (0D) models, taking only time and no spatial

© Springer Fachmedien Wiesbaden GmbH, part of Springer Nature 2019
M. Wentsch, *Analysis of Injection Processes in an Innovative 3D-CFD Tool for the Simulation of Internal Combustion Engines*, Wissenschaftliche Reihe Fahrzeugtechnik Universität Stuttgart, https://doi.org/10.1007/978-3-658-22167-6_2

dependence of the calculated parameters into account. Hence, no temperature and concentration distributions are considered. The changing combustion chamber conditions within one engine operating cycle are described by general valid differential equations for the mass and energy balance as well as the thermal equation of state for the working fluid [5, 67].

- **First law of thermodynamics** (conservation of energy)

$$Q_B + Q_W + H_I + H_E + W + H_L = 0 \qquad \text{eq. 2.1}$$

given the fuel heat release energy Q_B, wall heat transfer Q_W, intake enthalpy H_I, exhaust enthalpy H_E, leakage enthalpy H_L and work W. The differential notation of eq. 2.1 describes the change of internal energy U over time and allows a °CA-resolved analysis of an engine operating cycle:

$$\frac{dQ_B}{d\varphi} + \frac{dQ_W}{d\varphi} + \frac{dH_I}{d\varphi} + \frac{dH_E}{d\varphi} + \frac{dW}{d\varphi} + \frac{dH_L}{d\varphi} = \frac{dU}{d\varphi} \qquad \text{eq. 2.2}$$

- **General mass balance**

$$\frac{dm_C}{d\varphi} = \frac{dm_I}{d\varphi} + \frac{dm_E}{d\varphi} + \frac{dm_L}{d\varphi} + [\frac{dm_B}{d\varphi}]_{DI} \qquad \text{eq. 2.3}$$

where the change of cylinder mass m_C results from outflow of exhaust gas m_E and leakage mass m_L due to blow-by as well as inflow of fresh load m_I and fuel mass m_B in case of diesel or DISI-engines.

- **Thermal equation of state**

$$p \cdot V = m \cdot R_s \cdot T \qquad \text{eq. 2.4}$$

describing the relation between the thermal state variables of an ideal gas, i.e. pressure p, volume V, mass m, specific gas constant R_s and temperature T.

According to [16, 41], the working process calculation (WPC) enables a very fast cycle analysis and evaluation (less than 1s per revolution), but requires assumptions concerning the combustion profile. Very common are empirical combustion process models like the Vibe function. These static parametric functions need to be calibrated for each operating point. Their predictive capability can therefore be limited.

The thermodynamic modeling can be done for a one- or two- (or even multi) zone calculation, assuming a perfect homogeneity of temperature and gas composition within each zone. Both approaches are schematically illustrated in Figure 2.2. During combustion, the two-zone calculation additionally divides the cylinder filling, without spatial allocation of the zones, into a burned zone, containing exhaust gas with temperature T_b, and an unburned zone, which contains fresh charge and residual exhaust gas. The flame front is assumed to be infinitely thin and thermodynamically neglected. This provides information about the exhaust gas generation (in particular NO and CO) as well as the wall heat transfer, in contrary to the one-zone approach. Pressure p remains constant over the entire combustion chamber. A consideration of physical influences, e.g. cylinder geometry, flow or exhaust gas recirculation (EGR), on the combustion is not possible in either case. [5, 16, 41]

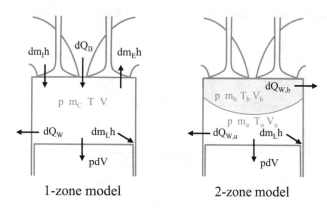

Figure 2.2: Combustion process modeling [41]

2.2 1D-CFD Simulation

In addition to a crank angle resolved temporal analysis of an internal combustion engine, the 1D-CFD simulation allows a spatial resolution of the engine along the main flow direction, as schematized in Figure 2.3. The simulation domain (incl. intake and exhaust system) is then divided into a finite number of sub-volumes and the flow field is calculated by means of conservation equations (continuity, momentum and energy) [16], which will be discussed in the next section. The cylinder is still modeled using the WPC and requires the definition of an empirical combustion profile. Alternatively, phenomenological or quasi-dimensional models, which consider influencing factors like the combustion chamber geometry or the in-cylinder flow motion (tumble, swirl), can be implemented. These combustion models, which are based on the two-zone or even multi-zone calculation, can adapt very well to changes in the operating conditions (e.g. engine speed, lambda, EGR, etc.), reducing the necessary calibration effort [5]. This is why they can be considered as predictive models, in contrary to pure empirical ones. Engine components with more complex geometries (e.g. turbocharger) are usually implemented via characteristic maps [16].

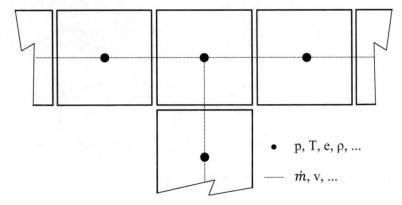

Figure 2.3: 1D-CFD discretization and calculation scheme [23]

1D-CFD simulations of internal combustion engines are characterized by acceptable calculation times due to a limited number of discretization elements (depending on the field of application). They are particularly convenient for the investigation and prediction of different engine strategies (exchange process, i.e. valve timings, turbo-charging layout, etc.) and transient engine operating conditions in order to optimize the entire engine map or, thanks to their versatility, the full vehicle behavior (including hybridized powertrains). 1D-CFD models allow a rapid definition of many engine parameters and general design features but their limit is reached when assessing geometry influences on the flow field, e.g. of the intake and exhaust channels, different combustion chamber or injector geometries. In particular injection processes are highly three-dimensional and cannot be sufficiently described by means of 1D-CFD simulations. [16, 41, 75]

2.3 3D-CFD Simulation

2.3.1 Fundamental Equations

3D-CFD simulations represent the most detailed and comprehensive approach to numerically investigate fluid dynamical problems. For this purpose, the domain of interest is discretized into a computational grid, consisting of a multitude of finite volumes (up to millions of cells) in which the reactive flow field is calculated. This is done on the basis of partial differential equations in dependence of time and three spatial coordinates. The coupling of conservation equations for mass, momentum and energy with an equation of state enables a thorough description of fluid flows inside an engine. A detailed derivation of the equations is not given at this point, since this is covered in numerous publications, e.g. [2, 15, 34, 41, 45, 65]. However an overview will be given in the following.

The differential notation of a conservation equation of an extensive quantity $F(t)$ can be derived from the integral balance at a fixed volume Ω with surface $\Delta\Omega$ [16, 65, 74], which in turn has a general valid Eulerian notation

$$\int_{\Omega} \frac{\partial}{\partial t} f \cdot \mathrm{d}V + \int_{\partial\Omega} \vec{\phi}_f \cdot \vec{n} \cdot \mathrm{d}S = \int_{\Omega} (s_f + c_f) \cdot \mathrm{d}V \qquad \text{eq. 2.5}$$

in which $f = f(\vec{x},t) = \mathrm{d}F/\mathrm{d}V$ with $f : \Omega \times (0,T) \to \mathbb{R}^3$ represents the density function of the conservation quantity (mass, momentum or energy). The temporal variation of the analyzed quantity within one control volume is captured by the first term of the equation. This is i.a. caused by the flow $\vec{\phi}_f \cdot \vec{n} \cdot dS$ (normalized space vector n) through the surface of the control volume, due to convection or diffusion processes. Here, the current density $\vec{\phi}_f : \partial\Omega \times (0,T) \rightarrow \mathbb{R}^3$ quantifies the amount of quantity F, which flows through the surface area per time unit.

The source terms s_f and c_f both describe changes of the conservation quantity, which can either be caused by local generation or reduction of the quantity inside the volume or globally by distant effects like gravitation or radiation. The application of the Gaussian integral theorem (divergence theorem) results in

$$\int_{\partial\Omega} \vec{\phi}_f \cdot \vec{n} \cdot dS = \int_{\Omega} div(\vec{\phi}_f) \cdot dV \qquad \text{eq. 2.6}$$

and therefore (based on eq. 2.5)

$$\int_{\Omega} \frac{\partial}{\partial t} f \cdot dV + \int_{\Omega} div(\vec{\phi}_f) \cdot dV = \int_{\Omega} (s_f + c_f) \cdot dV \qquad \text{eq. 2.7}$$

Differentiating with respect to the volume Ω yields the general valid differential notation of conservation equations, from which equations for mass, momentum and energy can then be derived

$$\frac{\partial f}{\partial t} + div(\vec{\phi}_f) = s_f + c_f \qquad \text{eq. 2.8}$$

Mass conservation equation (continuity equation)

The conservation of mass m is described by the continuity equation

$$\frac{\partial \rho}{\partial t} + div(\rho \vec{v}) = 0$$

$$f = \rho$$

$$\vec{\phi}_f = \rho \vec{v} \qquad \text{eq. 2.9}$$

Here, the density function $f(\vec{x}, t)$ is substituted by the mass density ρ and the mass flow can be described by building the product with the local flow velocity $\vec{v}$. Since the total mass does not change during the engine process, the source terms can be omitted ($s_f = 0$, $c_f = 0$).

If the distribution of mass fractions $w_i = m_i/m$ of individual species i within the flow region is of interest (e.g. fuel distribution inside the combustion chamber), the corresponding mass density $\rho_i = \rho \cdot w_i$ is applied. The local flow velocity $\vec{v}_i = \vec{v} + \vec{V}_i$ is given by the sum of the average flow velocity $\vec{v}$ and the diffusion rate $\vec{V}_i$, which generates the diffusion mass flow $\vec{J}_i$ of species i. The generation and reduction of the individual species, due to chemical reactions like combustion, can be expressed by means of the molar mass M_i as well as the molar formation rate ω_i. For the species mass, eq. 2.8 can therefore be reformulated as follows

$$\frac{\partial \rho w_i}{\partial t} + div(\rho w_i \vec{v}) + div(\vec{J}_i) = M_i \omega_i$$

$$f = \rho_i$$

$$\vec{\phi}_f = \rho_i \vec{v}_i = \rho_i(\vec{v} + \vec{V}_i) = \rho_i \vec{v} + \vec{J}_i \qquad \text{eq. 2.10}$$

$$s_f = M_i \omega_i$$

$$c_f = 0$$

Momentum conservation equation (Navier-Stokes equation)

In order to describe the motion of viscous fluids (e.g. liquid fuel droplets), the individual terms in eq. 2.8 can be adapted as follows. The momentum density $\rho \vec{v}$ is used as density f. The momentum flow is divided into a convective part

$\rho\vec{v}\otimes\vec{v}$ and a stress tensor of 2nd order $\bar{\bar{P}}$, with $\bar{\bar{P}} = p\bar{\bar{I}} + \bar{\bar{\Pi}}$. The latter describes a change of momentum due to the pressure p and viscous effects, which can be characterized by means of a shear stress tensor $\bar{\bar{\Pi}}$. Taking into account the gravitation $\rho\vec{g}$, the momentum conservation equation can be finalized.

$$\frac{\partial\rho\vec{v}}{\partial t} + div(\rho\vec{v}\otimes\vec{v}) + div(\bar{\bar{\Pi}}) - grad(p) = \rho\vec{g}$$

$$\begin{aligned} f &= \rho\vec{v} \\ \vec{\phi}_f &= \rho\vec{v}\otimes\vec{v} + \bar{\bar{P}} \\ s_f &= 0 \\ c_f &= \rho\vec{g} \end{aligned} \qquad \text{eq. 2.11}$$

Energy conservation equation

Conservation equations for the specific internal energy u and the specific enthalpy h, respectively, can be used to characterize the energy content of the fluid. The former can be derived from the total energy e formulation, which consists of internal energy, kinetic energy and potential gravitational energy G. The total energy density describes a corresponding composition

$$\rho e = \rho u + \frac{1}{2}\rho|\vec{v}|^2 + \rho G \qquad \text{eq. 2.12}$$

Further correlations for the total energy are as follows

$$\begin{aligned} f &= \rho e \\ \vec{\phi}_f &= \rho e\vec{v} + \bar{\bar{P}}\vec{v} + \vec{J}_q \\ s_f &= 0 \\ c_f &= q_r \end{aligned} \qquad \text{eq. 2.13}$$

According to that, the energy flow consists of a convective term $\rho e\vec{v}$, energy transport $\vec{J}_q$ (due to thermal conduction) and an additional term $\bar{\bar{P}}\vec{v}$, which characterizes the change of energy due to pressure and friction forces. In the source term c_f, effects q_r of radiation or magnetic fields are considered.

Kinetic and gravitational energy can be neglected at this point, since they only represent a minimum share of total energy in internal combustion engines. Thus, the conservation equation for the specific internal energy is

$$\frac{\partial \rho u}{\partial t} + div(\rho u \vec{v} + \vec{J_q}) + \bar{\bar{P}} : grad(\vec{v}) = q_r \qquad \text{eq. 2.14}$$

Considering the relation $\rho h = \rho u + p$ leads to the conservation equation for the specific enthalpy

$$\frac{\partial \rho h}{\partial t} - \frac{\partial p}{\partial t} + div(\rho h \vec{v} + \vec{J_q}) + \bar{\bar{P}} : grad(\vec{v}) - div(p\vec{v}) = q_r \qquad \text{eq. 2.15}$$

This transformation is reasonable, since the combustion chamber of an engine is an open system with mass flows and changing volume.

Internal engine flows can thus be described by means of a system of non-linear partial differential equations. Such a system can generally not be solved analytically, only by means of numerical solution methods. For this purpose, the above described differential conservation equations are converted into algebraic equations by discretization. Various methods are available for this purpose, e.g. the finite volume method. Here, the conservation equations are solved for each control volume within the computational mesh. These are represented by central computation points P (see Figure 2.4) and enable a calculation of all relevant flow parameters.

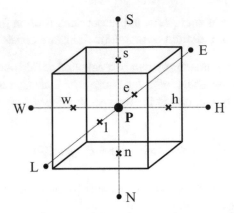

Figure 2.4: Central computation point P [16]

In order to transfer differential conservation equations into algebraic ones, they are integrated over the individual elements $\Omega(t)$ and evaluated subsequently. Using the Gaussian integral theorem, volume integrals are altered into surface integrals. This leads back to eq. 2.7. In addition, some estimations are necessary. $f(\vec{x},t)$, for example, can be assumed as homogeneous function over the volume element $\Omega(t)$ and the derivation can be substituted by its difference quotient

$$\frac{\partial}{\partial t} \int_{\Omega} f \cdot dV \approx \frac{f(\vec{x}_P, t + \Delta t) \cdot V_{\bar{C}}(t + \Delta t) - f(\vec{x}_P, t) \cdot V_{\bar{C}}(t)}{\Delta t} \qquad \text{eq. 2.16}$$

The flow term can be described by the arithmetic average of $f(\vec{x},t)$ and velocity $\vec{v}(\vec{x},t)$ over each sub-surface S_j $(\partial\Omega = \sum_j S_j)$ of volume Ω with normal vector $\vec{n}(\vec{x}_j,t)$.

$$\int_{\partial\Omega} \vec{\phi}_f \cdot \vec{n} \cdot dS = \sum_j (f(\vec{x}_j,t) \cdot \vec{v}(\vec{x}_j,t) \cdot \vec{n}(\vec{x}_j,t)) \cdot S_j \qquad \text{eq. 2.17}$$

where $j \in \{N,W,S,E,L,H\}$. This procedure, however, cannot be considered trivial and has to be examined critically. The quantity $f(\vec{x}_j,t)$ on the surface of volume element Ω can only be expressed by means of the central computation points. In the discretization approach of central differences, f is determined by linear interpolation of neighboring computation points. Usage of this type of approximation causes numerical oscillations in the solution. It is therefore recommended to use alternative methods, such as the upwind method. Here, the flow direction is taken into account and the value of quantity f corresponds to the closest upstream data point. More detailed explanations on this topic can be found, for example, in [45], chapter 4.2.

The third term of eq. 2.7, containing the source terms, can be approximated according to the first one:

$$\int_{\Omega} (s_f + c_f) \cdot dV = (s_f(\vec{x}_P, t) + c_f(\vec{x}_P, t)) \cdot \Omega \qquad \text{eq. 2.18}$$

The resulting system of algebraic equations can then be solved in every computation point, for each time step (temporal discretization) by means of selected numerical algorithms. The time steps Δt can be chosen individually, considering the Courant criterion (for more details see Chapter 6.3) and finding a reasonable compromise between level of detail and CPU-time. Together with the thermal equation of state, the internal engine flow can be fully described in terms of spatial and temporal resolved pressure and temperature fields.

2.3.2 Turbulence Modeling

The flow in an internal combustion engine is inherently turbulent. Therefore, the numerical calculation of turbulence is of utmost importance in 3D-CFD simulations. In order to solve the conservation equations described above, different turbulence model approaches are available. In Figure 2.5, these are schematized in terms of their essential characteristics. The direct numerical simulation (DNS) represents the most detailed approach in order to capture even the smallest turbulent length scales. Here, the equations are solved numerically and the implementation of turbulence models is completely renounced. Due to the therefor necessary high-resolution length and time scales, the utilization of DNS for the simulation of engine fluid dynamics with high Reynolds numbers is not acceptable [12, 16, 41]. Instead, it is a useful tool for fundamental research activities. This also applies to large eddy simulations (LES). Although sub-grid models are introduced for the calculation of the small-scale turbulence components, the "large eddies" are still explicitly determined, requiring highly refined computational meshes with millions of cells [16, 22, 61]. The standard for the simulation of engine flow fields and in particular for the purpose of virtual development is therefore the application of Reynolds averaged Navier Stokes (RANS) equations [41]. Here, the turbulent flow is completely described by models, which eliminate transient flow fluctuations via averaging. An explicit numerical solution of the equations is omitted. This results in a significant reduction of computational effort in comparison to DNS or LES and makes RANS rather applicable to large computational domains as well as high Reynolds numbers [58].

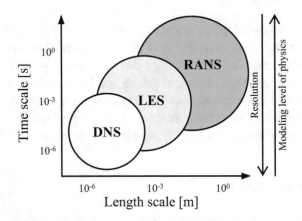

Figure 2.5: Modeling approaches for the description of a turbulent flow field

2.3.3 Application Possibilities and Limits

In theory, the application range of 3D-CFD simulations is widespread, providing high resolution results independent from the undertaken calibration effort. In reality, the computing time for a three-dimensional numerical flow field calculation is very high and strongly increases with the number of cells. Halving the cell edge length and therefore increasing the number of computational cells by factor 8, results in an up to 25 times higher CPU-time (central processing unit) [16]. A cell refinement usually requires yet an additional reduction of the time step increment in order to satisfy the Courant criterion. To still have reasonable CPU-times, it is then necessary to limit the analyzed engine domain (e.g. the combustion chamber, possibly only as half-mesh [40]). The extraction of engine components, however, requires a thorough definition of boundary conditions in order to represent the transition to the missing flow regions (e.g. the intake and exhaust system). These can either come from experimental measurements or 1D-CFD simulations. [16]

A well harmonized combination of a high quality calculation mesh (ideally manually generated hexahedral cells [41]), calculation models (e.g. turbulence, combustion, etc.) and accurate boundary conditions, enables a highly predictive and reliable simulation of complex thermo-fluid-dynamical phenomena [16]. This allows an investigation and evaluation of different operating strategies, design modifications or injection strategies. However, an immediate integration into the engine development process is only conditionally reasonable and feasible, due to the very high processing and computing time as well as the limited informative value of extracted simulation domains with respect to the full engine behavior (further details will be given in Chapter 3). Therefore, 3D-CFD simulations are more commonly deployed for research purposes or isolated (pre-) development tasks.

Especially in the investigation and evaluation of fuel injection processes, usually only the combustion chamber domain is considered. Its calculation mesh is then discretized increasingly fine towards the injector region. This proceeding results in millions of grid cells, which cause high CPU-times up to days for a single injection process, fully detached from the exchange and combustion process. The aim of this approach is to generate high-resolution information on the fuel exit and distribution. In some cases, this discretization procedure is taken so far that the cell edge length is chosen to be ten times smaller than the injector nozzle diameter, additionally incorporating an adaptive mesh structure for the expected spray propagation zone as illustrated in Figure 2.6.

Figure 2.6: Comparison of spray propagation using a regular mesh structure
and an adaptively refined mesh with injector nozzle resolution
[41]

According to the author, however, this must be regarded differentiated. If selected local fuel related phenomena are of interest, e.g. evolution of shock waves during gaseous fuel injection under supercritical conditions, such a procedure is reasonable. Since this leads to enormous computing times and possibly limits the predictive capability of a simulation (due to prior mesh adaption), this method is yet not very practical for time-limited project-specific engine development applications. Furthermore, the reliability of the results is limited by the fact that a definition of boundary conditions is necessary both for the initialization of the in-cylinder flow as well as the fuel injection itself. In particular, the latter cannot be measured readily and requires an additional simulation of the injector internal flow as well as the primary spray break-up [9, 41]. The high dependency on accurate boundary conditions can even counteract the demand for a high level of detail of the simulation results, initially intended by cell refinement.

3 The 3D-CFD Tool QuickSim

The scope and focus of various 3D-CFD codes are just as diverse as their application areas. This can be the examination of an isolated process within a single cylinder on the one hand and the investigation of gas dynamics within the intake system of a turbocharged 4-cylinder engine on the other hand [75]. The present work does not aim at comparing different approaches and assessing their quality regarding simulation results accuracy, level of detail, etc. Rather, the focus lies on a further development of the 3D-CFD tool QuickSim by means of extensive sensitivity analyses regarding the modeled fuel injection. QuickSim, its applicability and major functionalities are described extensively in [16]. However, since the procedure in the present work is essentially influenced by the specifics of this program, the main features as described in [16] will be discussed in the following.

3.1 The Purpose of QuickSim

Driven by the demand for a predictive and reliable 3D-CFD tool, which can be efficiently utilized within an internal combustion engine (ICE) development process, QuickSim was developed and is constantly enhanced at the Forschungsinstitut für Kraftfahrwesen und Fahrzeugmotoren Stuttgart (FKFS)/ Institut für Verbrennungsmotoren und Kraftfahrwesen (IVK), University of Stuttgart, starting in 1998.

Out of all (virtual) development tools, 3D-CFD simulations provide the most detailed information on an engine operating cycle. This includes spatially and temporally resolved flow fields, pressure and temperature distributions, chemical and thermodynamical processes, etc. In theory, this enables a thorough investigation of processes inside the engine and especially their interference with changes in geometry or operating strategies. As introduced before, a comprehensive application of 3D-CFD simulations seems inconvenient for in-

© Springer Fachmedien Wiesbaden GmbH, part of Springer Nature 2019
M. Wentsch, *Analysis of Injection Processes in an Innovative 3D-CFD Tool
for the Simulation of Internal Combustion Engines*, Wissenschaftliche Reihe
Fahrzeugtechnik Universität Stuttgart, https://doi.org/10.1007/978-3-658-22167-6_3

dustrial approaches, which exceed the pure research purpose. The informative value is usually limited to extracted engine domains (e.g. a single cylinder, optionally with parts of the intake or exhaust channels), selected time intervals within the engine operating cycle (e.g. from intake valve closing (IVC) to ignition point (IP)) or a few operating conditions. This is mainly due to extensive CPU-time demands but not practical, since the complexity of engine concepts and thus the variant diversity steadily increases. The setup (pre-processing), calibration, execution and evaluation (post-processing) of simulations can take several weeks, even using expensive high-performance hardware like super computers. In addition, qualified experts are needed for the simulation process, results analysis and interpretation, followed by a derivation and implementation of necessary changes in order to improve the engine concept. The reliability and predictability of calculations can be very limited by a lack of understanding the occurring physical phenomena and their incorrect mathematical description [34]. Dependencies of the calculation models on the mesh structure as well as wrong assumptions on initial and boundary conditions can moreover lower the results quality.

The development of QuickSim as a solution-oriented 3D-CFD tool, which uses the commercial CFD code STAR-CD in the background, was driven by the aim to provide solutions for these application restricting factors in order to offer financial and temporal benefits and enable comprehensive studies and a holistic virtual development of internal combustion engines (even during the prototype phase) [64]. In concrete terms, several development objectives were formulated, whereof three will be introduced here:

Fast Calculation

In order to enable applicability within constantly shorter development cycles, the simulation process chain as well as the actual numerical computation needs to be significantly accelerated. To realize this without additional computing power, an application dependent reduction in the number of cells is required, without sacrificing the quality of the simulation results.

Only time savings over more conventional 3D-CFD approaches enable the implementation of QuickSim as a powerful development tool, which allows various engine concepts to be calculated, operating characteristics to be directly analyzed and changes in the geometry or control strategies to be rapidly realized. [75]

Reliability of Results

An increase in reliability, as a prerequisite for meaningful and practice-oriented simulation results, demands a holistic examination of the thermo- and fluid-dynamic system and a high degree of predictive capabilities. For this reason, potentially negative influencing factors must be minimized. These include, for example, the computational grid, user-defined initial and boundary conditions as well as implemented calculation models.

This significantly reduces the calibration and validation effort on the basis of experimental data and supports a gradual independence from test bench measurements. Moreover, it enables the utilization of QuickSim even before the first prototype setup (gas-related predesign of the internal combustion engine: gas exchange, mixture formation, combustion). This consequently increases the development process efficiency, due to cost (less component manufacturing and measuring campaigns) and additional time savings.

Clear Results Presentation

3D-CFD simulations are well-known for their colored contour plots or spray droplet images. Further information on the engine behavior is usually hardly accessible and rarely processed for development engineers. In order to reasonably integrate a 3D-CFD tool into the development process, data must be provided in such a way that objective concept decisions can be made. In addition, they should allow a direct comparison with the results of other simulation tools or experimental measurements.

Some essential features of QuickSim, which are necessary to meet the priorly listed requirements, will be discussed in the following.

3.2 Features of QuickSim

3.2.1 Fast Analysis 3D-CFD Tool

As already indicated, the number of discretized volumes, and thus the number of equations to be solved, represents the key influencing factor on the duration of numerical calculations. An arbitrary cell coarsening, without adaption of the computational models, however, results in a distinct distortion of the simulation outcome.

The calculation models implemented in QuickSim represent a well-considered combination of traditional local 3D-CFD models (for example viscosity of air, droplet wall impingement, etc.), engine-specific phenomenological relationships, trained neural networks, databases and, if needed, empirical relationships. Specifically ICE-adapted models, which consider cell dimensions as well as the mesh structure, do then not have a general thermodynamic validity anymore. However, application-oriented modifications in favor of the CPU-time are entirely justifiable for the purpose of virtual engine development, e.g. by resigning detailed chemical reaction kinetics (additionally describing the working fluid composition by means of only six species scalars), while maintaining the necessary thermodynamic accuracy of the fuel heat release and wall heat transfer (internal check of plausibility by means of integrated real working process calculation), also resulting in specific benefits such as the availability of real fuel models [75]. A realistic flame propagation can furthermore be calculated by means of a localized (cell specific) phenomenological two-zone approach.

Depending on the geometrical complexity of the combustion chamber, these adaptions and optimizations enable a reduction of cells to approx. 30,000 to 50,000 cells per cylinder (instead of min. 1,000,000 [41]), which corresponds to an average cell discretization length of approximately 2.5 mm. Employing an average time step of 0.5 °CA, this results in an overall decrease of CPU-time up to a factor 100 (compare Figure 3.1) without sacrificing accuracy of results, using a single state-of-the-art processor.

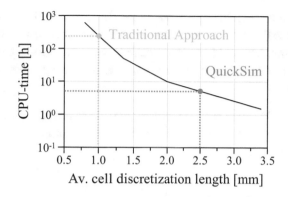

Figure 3.1: CPU-time of one engine operating cycle in dependence of the cell discretization length (single cylinder incl. parts of intake and exhaust channels) [16]

Within a logarithmic time scale, as represented in Figure 3.2, the 3D-CFD simulation with QuickSim can therefore be classified differently than conventional 3D-CFD simulations.

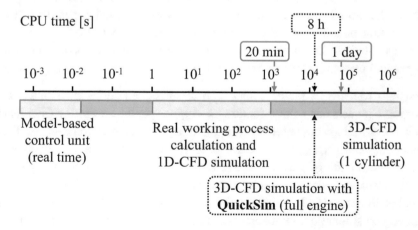

Figure 3.2: CPU-time classification of QuickSim

A parallelization on multiple processors would enable further savings in calculation time (ideally a linear reduction with the number of processors). However, the advantage would become obsolete with increasing number of cells and processors due to the necessary communication between the individual processors (max. 6).

Thanks to the CPU-time reduction, it is possible to analyze more engine component geometries and control strategies in order to completely exploit the full potential of the engine. [75]

3.2.2 Full-Engine 3D-CFD Simulation

A simulation is ideally a replica of reality within a limited spatial domain [25]. The predictive capability of 3D-CFD simulations is therefore directly linked to the quality of defined boundary conditions. The end of an extracted engine flow domain (see Figure 3.3) ideally requires 3D-resolved pressure and temperature distributions as input in order to derive the correct intake/exhaust mass flow. Three-dimensionally, these can be neither delivered by 1D-CFD simulations nor by experimental measurements. The definition of locally averaged 1D boundary conditions should therefore be located at a preferably uncritical flow positions, minimizing their (negative) influence on the in-cylinder flow calculation. The problem here is in particular the consideration of time-dependent pressure waves in the local flow.

Starting from the extracted combustion chamber, an addition of intake and exhaust channels can neither capture the engine specific air motion (incl. backflow phenomena) nor the fuel distribution of a port fuel injection concept. Even appending an airbox geometry does not seem sufficient, since it results in even more "mesh endings", which require additional definition of boundary conditions. Here, cylinder-to-cylinder variations cannot be determined. These examples, illustrated in Figure 3.4 point out the necessity of a holistic 3D-CFD full engine simulation, including all cylinders, ports, channels, airbox, etc., to realistically reproduce the engine behavior.

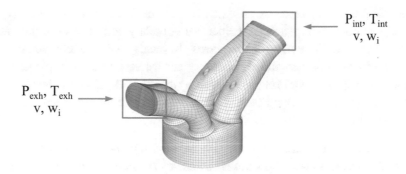

Figure 3.3: Definition of boundary conditions at intake and exhaust mesh
endings

Due to the model-related time-savings, QuickSim enables the simulation of
the entire engine domain. The exemplified engine model in Figure 3.4, starts
from the throttle position and includes the flow domain up to the turbine inlet
and wastegate. Such an engine mesh can be calculated within acceptable CPU-
times, i.e. approximately 6-8 hours per operating cycle (on a single processor)
for a 4-cylinder spark-ignition (SI) engine.

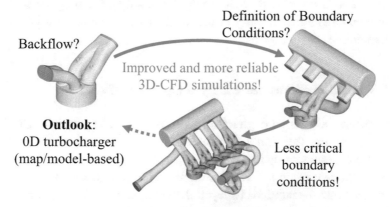

Figure 3.4: 3D-CFD simulation domain extension

The ideal case, however, is an extension of the domain up to ambient conditions at both ends, since the respective boundary conditions are then atmospheric ones and free of pressure waves. In the case of turbocharged engines, an additional improvement of reliability can be achieved by implementation of a 0D (map or model based) or even a (decoupled) 3D turbocharger, virtually closing the relevant air path. Both are currently being implemented in the QuickSim code.

An extension of the simulation domain and therefore the definition of boundary conditions at less critical positions, enhances not only the numerical stability but also significantly influences the accuracy and therefore the predictive capability of the simulation, permitting a detailed virtual design of the internal combustion engine. During the engine development process, critical operating characteristics can be identified (also for individual cylinders) and countermeasures can be taken, e.g. by modifying the airbox or channel designs. [75]

Depending on the engine domain setup, it is mandatory to thoroughly calibrate the engine simulation for one configuration, ideally by means of experimentally determined data. Main calibration quantities are the total air mass flow, which can be regulated by the defined intake pressure and temperature, as well as correcting variables for the combustion process. Based on a well calibrated model it is then possible to simulate further configurations (geometries and operating points) independently of the test bench and without further need of calibration.

3.2.3 Simulation of Successive Engine Operating Cycles

To reproduce the internal engine flow field as realistic as possible, QuickSim also benefits from the ability of simulating more than one single operating cycle. Here, results of cycle n are used for the initialization and calculation of the subsequent cycle $n + 1$. This significantly minimizes the influence of user-defined initial conditions for cycle 1 (temperature, pressure, lambda, etc.) and a stable holistic flow field can evolve. As indicated in Figure 3.5, the duration to achieve convergence of results (primarily the air mass flow) highly depends on the calculation mesh size. In particular the simulation of a full engine requires the calculation of several successive engine operating cycles in order to ensure convergence (after approx. 6 to 8 cycles for a 4-cylinder SI-engine). The maximum number of cycles is theoretically unlimited. The only restrictive factors are the maximum available computation time as well as storage capacity.

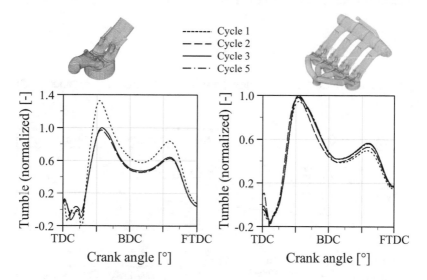

Figure 3.5: Comparison of multi-cycle simulations for single-cylinder and full engine domain (cylinder 3; 5500 rpm, WOT; non-dimensional) [16]

In addition to simulating steady state operating conditions, the multitude of cycles even allows the simulation of small transient state changes, e.g. by changing the injection or ignition timing from one cycle to another. Furthermore, cycle-to-cycle variations, which are caused by emerging flow fluctuations, can be captured very well. However, cyclic deviations in the spark behavior or similar can not be determined, due to their control by means of fixed input variables (e.g. spark duration, spark kernel size, etc.).

The flow field convergence has a particularly high influence on the charge motion within the cylinder and thus on the mixture formation (compare Figure 1.1). In order to reliably investigate engine specific injection processes,

including the droplet breakup (more details in Chapter 4.3) and fuel distribution, it is therefore reasonable to deploy full engine simulations over successive working cycles. For multi-component fuel simulations, which will be
discussed in Chapter 7, it suffices to start the simulation from an already converged operating cycle in order to save CPU-time.

3.2.4 Information Exchange Layout

Figure 3.6 schematically illustrates the information exchange layout of Quick-
Sim.

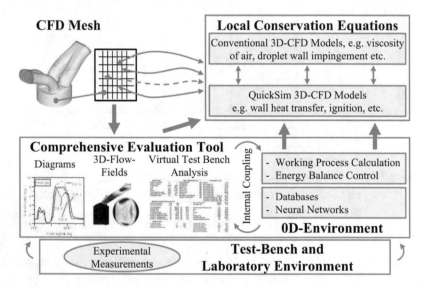

Figure 3.6: Information exchange layout of QuickSim [16]

The essential basis of simulations with QuickSim represents a high-quality
3D-CFD mesh. This is created half-automated with hexahedral cells inside
the combustion chamber, channels and the exhaust manifolds. For the airbox
unit it can also be reasonable to use polyhedral instead of hexahedral cells (depending on the geometry and operating conditions). Due to the implemented
models, an additional cell refinement in the vicinity of the combustion chamber

walls is not necessary. When preparing the mesh, it is of utmost importance that no undesirable distortions, e.g. concave cell structures, occur during the mesh motion generation.

As described in Chapter 2, the conservation equations are solved locally in each cell's computation point. The application of the thermal equation of state together with implemented 3D-CFD models, e.g. for the combustion, heat transfer, turbulence and others, partly using input from databases and neural networks, enables a description of the changing combustion chamber conditions within each time step. An integrated real working process calculation serves as internal control and correction function. Global 3D-CFD simulation results (e.g. fuel heat release, wall heat transfer, internal energy variations, etc.) are compared to the simultaneous running real working process calculation at each time step. Deviations are locally corrected by means of a global correction factor, ensuring a high level of accuracy in the solution of relevant engine processes.

A clear and detailed presentation of the simulation results, without ambiguity, is provided by a comprehensive evaluation tool. This includes graphics and animations, diagrams, tables and a virtual test bench analysis, offering an overview of all relevant engine parameters. In this way, an immediate comparison with experimental data, which is not only used as simulation input but in particular as reference data for results validation, is possible. Development engineers can herewith rapidly interpret and evaluate the analyzed engine configurations and take appropriate optimization measures.

3.3 Development History and Portfolio of QuickSim

The idea for the development of QuickSim first emerged in 1997, during the preparation of a master thesis about the simulation of the charge exchange process, using a conventional 3D-CFD program [Marco Chiodi, personal communication, November 22, 2016]. The actual programming of QuickSim started in 1998, first enabling a cylinder simulation with an integrated evaluation tool and including a basic 3D-CFD model development. Over the following years, these models were constantly improved and adapted to new applications. QuickSim version 3 was able to simulate an extended simulation domain, including the airbox geometry. From then on, not only the simulation domain was constantly expanded (first full engine simulation of a turbocharged 4-cylinder CNG engine in 2009) but also the application fields of QuickSim simulations, e.g. 2-stroke engines, HCCI combustion and many more.

According to the current state, the QuickSim portfolio can be summarized as follows:

- **Engine layout**
  No restrictions concerning the engine design, i.e. any cylinder number, combustion chamber geometry, intake and exhaust system geometry and injection system (direct injection, port fuel injection (single-point injection, multi-point injection, etc.)) with arbitrary injector geometry can be realized.
- **Ignition type**
  Any ignition type can be realized, i.e. spark ignition, compression ignition, HCCI.
- **Fuel type**
  Models for any fuel type can be implemented, e.g. gasoline, diesel fuel, compressed natural gas (CNG), bio fuels, synthetic fuels.
- **Operating strategy**
  Variable valve and piston motion; any injection strategies.

3.4 Status Quo of Injection Modeling with QuickSim

It has already been emphasized, that the definition of boundary conditions and model input parameters has a considerable effect on the significance of the simulation output. In order to keep this influence as low as possible, Quick-Sim can perform i.a. full-engine simulations over several successive operating cycles, as explained in Chapter 3.2. Similar measures concerning the boundary conditions and input parameters for the fuel injection with QuickSim have not been taken to a large extent.

At the beginning of the research activities presented in this work, the fuel injection description was very basic, mostly using the standard models and subroutines (droico.f, dropro.f, drmast.f) of STAR-CD. Although more details will be given in the following chapters, the status quo at that time is to be briefly presented by means of the classification introduced in Figure 1.2, comprising the fuel modeling, injector and injection strategy implementation as well as numerical framework conditions.

Fuel modeling

Both, the injection of liquid and of gaseous fuels could be realized. Here, the fuel was represented by means of a single-component model (according to requirements, also utilizing substitutes like n-heptane, iso-octane or methane). Its liquid properties, which govern the breakup, evaporation and distribution processes, were defined by constant values (for density, viscosity, pressure of saturation, etc.) and an averaged $C_nH_mO_rN_q$ assignment, on the basis of which also the fuel caloric database is generated.

Injector and injection strategy

The injector geometry and position definition was already highly flexible, enabling the integration of any injector type (single-hole, multi-hole, hollowcone, etc.) inside the combustion chamber, intake channels (MPI) or the airbox inlet (SPI). The geometrical layout was usually directly adopted from the spray targeting provided by the injector manufacturer. Spray angle variations during the injector opening and closing phase have been neglected.

Input parameters in dependence of the injection conditions, i.e. injection pressure, temperature and others, had to be estimated in the absence of reference data. These include, for example, the initial droplet velocity and droplet size distribution.

Numerical framework

The influence of numerical boundary conditions on the process of fuel injection, i.e. calculation mesh refinement, temporal discretization as well as convergence criteria, has not been explicitly investigated.

3.5 Research Objective

Based on the explicated status quo, the aim of this work can be summarized in answers to the following questions:

- How detailed does the injection have to be modeled?

- How to find appropriate injection boundary conditions?

- Do specific boundary conditions have a general validity, regardless of their application?

- How sensitive is the injection modeling to changes in the boundary conditions?

- How can the information content of the injection simulation be enhanced without significantly increasing the computation time?

4 Fundamentals of Spray Modeling and Simulation

In preparation of the analyses discussed in Chapters 6 to 8, an introduction to the modeling of injection sub-processes as well as required terminologies and calculation formulas will be given in the following. This includes fuel properties, spray atomization processes and the numerical injection definition as influencing parameters on the spray development and mixture formation (compare Figure 1.2).

A detailed description of, for example, fuel chemistry and composition as well as the exact functionality of different injector types will be omitted here. More details on this can be found in the relevant technical literature, e.g. [11, 18, 26, 66, 71, 73].

4.1 Fuel Properties and Specifications

Analyses, presented in this work, were mostly carried out with gasoline and CNG. This is why only these fuels are specifically described in the following. However, it should be emphasized that there are no restrictions with QuickSim concerning the choice of fuel type.

4.1.1 Liquid Fuel - Gasoline

The properties and behavior of liquid fuels are decisively influenced by their composition. Gasoline consists of hundreds of hydrocarbon compounds, which can be divided into chemical groups according to the arrangement and nature of their bonds [4, 40]. The different molecular sizes and structures essentially determine the thermo-physical characteristics of the components. Their composition, in turn, identifies those of the fuel.

© Springer Fachmedien Wiesbaden GmbH, part of Springer Nature 2019
M. Wentsch, *Analysis of Injection Processes in an Innovative 3D-CFD Tool for the Simulation of Internal Combustion Engines*, Wissenschaftliche Reihe Fahrzeugtechnik Universität Stuttgart, https://doi.org/10.1007/978-3-658-22167-6_4

The fuel properties, considered in the present work, to describe the breakup and evaporation behavior of liquid fuel, include the density ρ, viscosity η, surface tension σ, specific heat capacity c_p, heat of vaporization h_v, saturation pressure p_s and boiling temperature T_b. These are, except for the latter, highly temperature dependent quantities. According to [73], the pressure dependency can be generally neglected, except for the fuel density. Within the presented analyses, however, the pressure dependency was only taken into account for gaseous fuels.

[73] provides comprehensive empirically determined data on the aforementioned properties for a multitude of substances. This includes correspondingly derived equations, which are listed in Appendix A2 (eq. A2.1 to eq. A2.6) to describe the thermo-physical behavior as a function of temperature, as well as the fluid-specific input-parameters $(A - E)$, caloric and critical data (T_c, p_c). On the basis of these correlations, it is possible to create an extensive database for pure substances. This, in turn, generates QuickSim-compliant, application-specific fuel input files, which provide the fuel data to the ICE simulation at any time and for any temperature conditions.

4.1.2 Gaseous Fuel - Compressed Natural Gas

CNG has methane as its main component (85-99 % max. content), which decisively determines the fuel characteristics. Its favorable H/C-ratio in comparison to gasoline* results in a considerable reduction of CO_2 emissions. Additional benefits like potential CO and NO_x reduction, higher knock resistance, etc. make CNG a promising alternative fuel. [10, 70]

This is why injection concepts with CNG are being increasingly investigated by means of 3D-CFD simulations, which provide not least an optical access to the occurring complex phenomena. In particular, the direct injection of CNG is moving into focus of research, since the currently prevailing port fuel injection causes a loss in the volumetric efficiency of approximately 8 % compared to gasoline with identical engine efficiency, due to the fuel's low density [10, 28]. However, the development of a production-ready CNG-DI strategy is

---

*H/C-ratio of methane: 4/1; H/C-ratio of gasoline: 1.75/1–1.9/1

a nontrivial task - neither the choice of well targeted injection parameters (e.g. injection pressure, start of injection (SOI), injection direction, jet shape, etc.) nor the numerical description of gaseous injection under supercritical conditions and the subsequent mixing process. [62, 63, 64, 76]

Due to the dominant methane content, CNG is simulated as a single-component model. Moreover, it is common practice to utilize the reference fuel G20 for experimental measurement, which consists of approximately 99.5 % methane. Its properties are also available in [73]. The corresponding equations for the temperature-dependent gaseous fuel data (dynamic viscosity, specific heat capacity) can be found in Appendix A2 as well. The density, in turn, is derived from the thermal equation of state (eq. 2.4). Properties like the surface tension, heat of vaporization and pressure of saturation are apparently irrelevant for gaseous fuel.

4.2 Numerical Injection Definition

The injection simulation with QuickSim does not physically integrate an injector into the computational mesh. Instead, a coordinate system is defined describing its mounting position, where the z-axis represents the injector middle axis oriented towards the combustion chamber. In case of an incisive injector tip, a rough geometry can be embedded within the cell structure, as exemplified in Figure 4.1.

The parametric description on the basis of four characteristic spray angles, allows to define a variety of injector geometries. As schematically illustrated in Figure 4.2, the spray jet middle axis orientation of a multi-hole injector is defined by spherical coordinates. The zenith angle (or polar angle) $\theta \in \{0, 180\}$ describes the radial distance from the injector middle axis z. Starting from the reference direction (x-axis), the jet orientation is finally determined via the azimuth angle $\varphi \in \{-180, 180\}$ or $\{0, 360\}$. Outgoing from the jet middle axis, the jet width is defined by means of angle α.

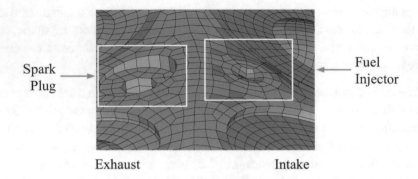

Spark
Plug →

← Fuel
Injector

Exhaust   Intake

Figure 4.1: 3D-CFD mesh including spark plug and injector tip geometry

Since the middle axis of an outward-opening injector jet coincides with the injector middle axis z, θ and φ equal zero. α is then identical to the half spray cone angle ε and the hollowcone structure is represented by an additional angle $\gamma \leq \alpha$. Except for γ, the geometric angles can generally be taken from the spray targeting, provided by the injector manufacturer. Table 4.1 exemplifies injector type specific angle definitions.

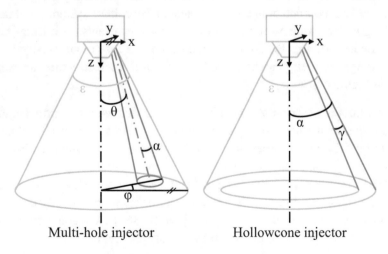

Multi-hole injector   Hollowcone injector

Figure 4.2: Spray angle definition

Table 4.1: Examples for injector type specific geometry parameters

Type	zenith θ	azimuth φ	α	γ
Outward-opening injector (hollowcone)	0°	0°	45°	3°
Single-hole injector	0°	0°	50°	50°
Multi-hole injector (e.g. regular 6 holes)	40°	0°, 60°, 120°, 180°, 240°, 300°	11°	11°

4.2.1 Liquid Fuel Injection

The injection of liquid fuel is modeled using a well-defined droplet initialization. In doing so, neither the injector internal flow nor the primary droplet breakup are directly calculated. In order to not cause an unacceptable high computational time, the concept of parcels is introduced according to the "Langrangian discrete droplet method": on the basis of statistical assumptions, a defined number of non-interacting droplets, which have identical physical properties, is numerically bundled into samples, so-called parcels. The number of parcels needs to be large enough so that a sufficiently accurate representation of the totality of droplets can be guaranteed (compare analyses in Chapter 6.1.3). [15, 27, 41]

For the reason of clarity, droplets and parcels are subsequently referred to as droplets, unless a strict distinction is necessary.

The liquid fuel droplets will be introduced in a limited conical spray region near the injector orifice. This region is delimited by α and γ as well as a minimum and maximum axial distance ($L_{\min}$ and $L_{\max}$) from the origin of the injector coordinate system (compare Figure 4.3). This procedure makes it not only possible to define the injector orifice position but also ensures numerical stability, since the droplet initialization is not limited to very few boundary cells [16].

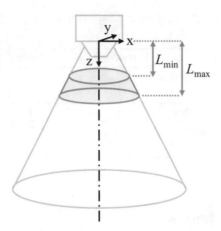

Figure 4.3: Definition of L_{min} and L_{max}

Initial values for position, size, velocity, direction of movement and temperature are assigned to each droplet [15] in consideration of the (user-defined) injection setup. Aside from the injector geometry this includes i.a. the following input parameters:

- **Mass flow rate** $\dot{m} = \mathrm{d}m/\mathrm{d}t$ [mg/s]:
  The flow of mass m through a section per unit time t can be determined at the test bench or calculated by

$$\dot{m} = \rho \cdot v \cdot A \qquad\qquad \text{eq. 4.1}$$

and depends on the density ρ of the fuel, the cross-section A of the injector orifice as well as the average flow velocity v.

Specifications concerning $\dot{m}_{max}$ are included in the spray targeting, usually based on measurements with n-heptane at an injection pressure of 10 MPa. By means of the Bernoulli equation (and the discharge velocity derived therefrom), the mass flow rate can be adapted to the prevalent pressure conditions. In addition, the changing cross-section during the injector opening and closing phase can be taken into account.

- **Sauter mean diameter (SMD)** D_{32} [μm]:
  The SMD is a characteristic parameter to describe the droplet size distribution within a fuel spray and represents the diameter of the averaged volume-surface-ratio of all injected droplets. According to [26], the SMD has gained the greatest importance as a reference parameter and can best describe the conditions within the fuel spray.

  As simulation input parameter it can either be acquired by means of PDA (phase doppler analyzer) measurements or needs to be estimated in consideration of the injector geometry as well as the injection pressure. An implemented Rosin-Rammler distribution finally employs the SMD to initialize a statistically representative droplet spectrum into the combustion chamber.

- **Injection velocity** v [m/s]:
  The specification of an averaged injection velocity in dependence of the injection pressure allows the assignment of an initial velocity to each droplet.

  v can be determined via PDA measurements as well. Alternatively, a theoretical calculation under the assumption of an ideal flow (incompressible fluid, no friction) is feasible.

  $$v = \sqrt{\frac{2 \cdot \Delta p}{\rho}} \qquad \text{eq. 4.2}$$

  where $\Delta p = (p_{inj} - p)$ is the difference between the injection pressure p_{inj} and the pressure p in the nozzle vicinity, e.g. inside the combustion chamber.

In the standard Euler-Lagrange formulation to describe the multi-phase flow, which is schematically illustrated in Figure 4.4, the individual droplet properties (including the fuel-specific characteristics) govern the secondary breakup as well as the phase exchange process for mass (evaporation), momentum and energy.

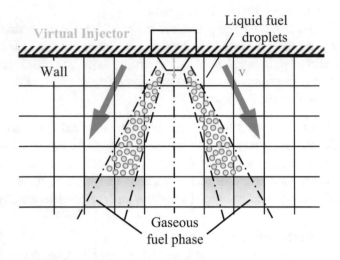

Figure 4.4: Liquid fuel injection modeling

The Eulerian approach is used for the calculation of the continuous gas phase. Here, the fluid motion is determined by means of fixed control volumes using the conservation equations defined in Chapter 2.3. For the calculation of the dispersed droplet phase, differential equations are solved in a moving coordinate system. This Lagrange method provides independent trajectories for each droplet, whereby their position is known at any time. The continuous and disperse phase influence one another in their movement, due to the transition of mass, momentum and thermal energy. This is controlled by source terms, which are in turn involved in the conservation equations. In this way, processes like the droplet breakup, distribution, collision, coalescence and wall interaction can be described. For the present work, however, the droplet-droplet interactions are neglected and the respective models have been disabled. [15, 26]

4.2.2 Gaseous Fuel Injection

Modeling of gaseous fuel injection is to some extents similar to that of liquid fuel but still possesses some peculiarities in the numerical description. The gaseous fuel is initialized by means of fictive gas droplets, following the above described procedure of liquid fuel injection. Within the first iteration, an immediate fictive evaporation takes place, free of evaporation enthalpy [16]. This procedure is schematically illustrated in Figure 4.5.

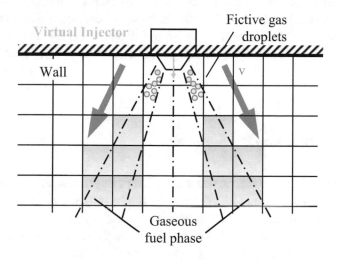

Figure 4.5: Gaseous fuel injection modeling [16]

This modeling approach has some limits in its applicability. It is, for example, not able to directly simulate supercritical injection conditions, which result in a supersonic gas jet, including its under-expanded regions and high Mach numbers. Such an approach would require a highly resolved computational mesh (and therewith smaller calculation time increments), having a cell discretization length that is multiple times smaller than the cross-sectional width of the injector orifice, implemented for example in [3, 41, 52, 58]. As described in Chapter 3, this is not compatible with the principles of QuickSim and according to [16] not necessarily required for the purpose of a virtual engine development. However, dispensing with a detailed gas spray makes a thorough injector calibration essential. In order to still capture the gaseous fuel propagation (i.a. penetration) within the combustion chamber, special attention needs to be paid to an appropriate parametric description of the expanded fuel jet (e.g. its density, initial velocity, etc.). Here, the Lagrangian approach enables to reproduce high velocities in the nozzle vicinity without significantly influencing the physical processes. Alternative approaches have been tested, e.g. the definition of an inlet boundary, directly initializing the gaseous fuel. This, however, failed due to two main reasons: the lack of reasonable 3-dimensional boundary conditions at the injector orifice, which are indispensable if the injector internal flow is not simulated, as well as the inability of capturing the actual spray penetration. An initialization of fictive gas droplets, on the contrary, allows the reproduction of the high fuel momentum caused by supercritical injection conditions.

4.3 Spray Breakup Mechanisms

The breakup of a liquid jet can occur in different ways, primarily depending on the discharge velocity and its influence on the (aerodynamic) forces affecting the jet. In the case of fuel injection and the demand for a breakup commencing at the nozzle exit as well as preferably small droplets, the regime of atomization, which was first introduced by [54], is required [26, 71]. As soon as the liquid jet exits the nozzle, its stability depends on the relation of the surface tension, the internal forces of the jet and the destabilizing forces induced by the interaction with the surrounding medium [48, 71]. In dependence of these forces, the breakup can be physically described by means of dimensionless quantities [26, 48]:

- **Weber number** We
  In order to make statements about the breakup of fluids with low viscosity, the Weber number serves as characterizing quantity, putting inertia forces of a fuel droplet in relation to its surface tension σ_f.

$$\mathrm{We} = \frac{\rho_g \cdot d_D \cdot v_{rel}^2}{\sigma_f} \qquad \text{eq. 4.3}$$

Here, the inertia forces affecting the droplet incorporate the density ρ_g of the surrounding gaseous fluid, the droplet diameter d_D (as problem specific characteristic length) and its relative velocity v_{rel}.

- **Reynolds number** Re
  In addition to the surface tension, also the dynamic viscosity η_f of the liquid can serve as stabilizing factor counteracting the breakup. The Reynolds number indicates the ratio of inertia and viscous forces.

$$Re = \frac{\rho_f \cdot d_D \cdot v_{rel}}{\eta_f} \qquad \text{eq. 4.4}$$

- **Ohnesorge number** Oh
  The Ohnesorge number reflects the relation of the viscosity to the surface tension of the fluid and is therefore determined by the fuel characteristics.

$$Oh = \frac{\eta_f}{\sqrt{\sigma_f \cdot \rho_f \cdot d_D}} = \frac{\sqrt{We}}{Re} \qquad \text{eq. 4.5}$$

The "Ohnesorge diagram" in Figure 4.6 shows the different spray breakup regimes, including the atomization regime, as a function of the Reynolds and Ohnesorge number. [48] underlines that under the assumption of a constant Ohnesorge number, i.e. consistent fuel properties and injector geometry, an increase of the discharge velocity (due to higher injection pressure) and therefore of the Reynolds number, leads to a higher order of spray breakup.

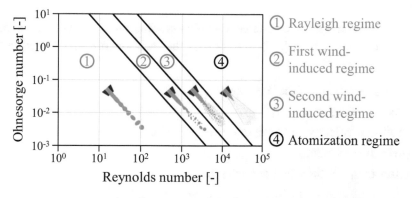

Figure 4.6: Spray breakup regimes (Ohnesorge diagram) [14, 48]

Fuel spray atomization includes a multitude of individual, partially simultan-
eously occurring mechanisms, which are depicted in Figure 4.7. The two main
mechanisms, primary droplet breakup and secondary or aerodynamic droplet
breakup can be well characterized by means of the afore introduced dimensionless quantities and will be explained in more detail.

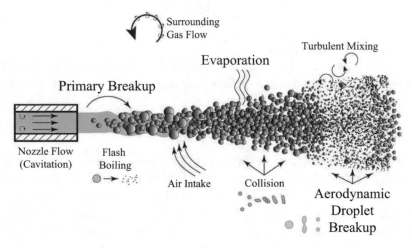

Figure 4.7: Spray atomization mechanisms [38]

The jet atomization is considered a highly complex process due to the numerous influencing factors [54], which cause the mentioned forces to act on the
fuel droplets:

- Nozzle geometry of the fuel injector,
- fuel specifications,
- discharge velocity and level of turbulence within the fuel jet,
- pressure conditions,
- temperature of the ambient medium.

Primary Droplet Breakup

The primary droplet breakup describes the formation of ligaments and droplets from a coherent liquid phase. This process occurs close to the injector orifice and is governed by the jet velocity and nozzle flow turbulence. In the case of injector internal cavitation (in particular for high pressure Diesel injection), the primary breakup already takes place inside the nozzle. Depending on the conditions, droplets of different size and shape are formed. [41, 71]

Further explanations will be omitted here, since the primary droplet breakup (as well as the injector internal flow) is not directly included in the QuickSim injection modeling. However, when omitting the primary droplet breakup, [41] recommends to thoroughly calibrate the secondary droplet breakup parameters as well as the initial droplet size by means of optical spray measurements. Varying temperature conditions in an injection chamber result in a deviation of the liquid spray penetration depths. An accurate choice of modeling parameters should be able to fully capture these temperature-dependent variations.

Secondary Droplet Breakup

The secondary breakup of liquid fuel droplets, which by tendency occurs further downstream the nozzle, is predominantly influenced by aerodynamic interactions between the dispersed droplet phase and the surrounding continuous gas phase [27, 71]. A distinction of different mechanisms is based on characteristic Weber number values. [49] provides a detailed classification of the secondary breakup, which is illustrated in Figure 4.8.

The standard Reitz-Diwakar model, which is implemented in STAR-CD and therefore utilized by QuickSim, only differs between two main types of secondary breakup [15]:

- **Bag breakup**
  Irregular pressure conditions surrounding the droplet cause its expansion. If the acting forces thereby exceed the surface tension of the droplet, breakup into many smaller droplets occurs.

- **Stripping breakup**
  The droplet breakup is caused by constant detachment or stripping of liquid
  from the droplet surface.

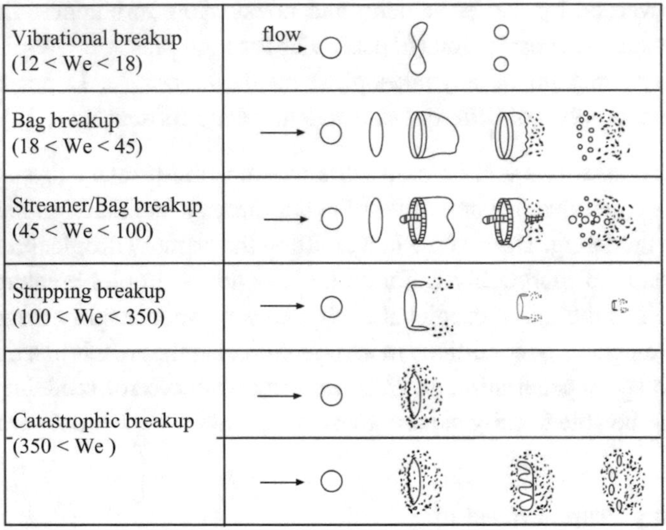

Figure 4.8: Secondary breakup mechanisms according to Pilch and Erdmann
          [49]

Employed calculation formulas, explicated in [15] do not only determine the
instability mechanism but also the corresponding droplet breakup rate within
a characteristic breakup time τ. Here the droplet diameter d_D is reduced until
reaching a stable droplet size $d_{D,s}$:

$$\frac{\mathrm{d}d_D}{\mathrm{d}\tau} = -\frac{(d_D - d_{D,s})}{\tau} \qquad \text{eq. 4.6}$$

Just like the characteristic time scale also the stable droplet size depends on
the breakup mechanism. It can be determined by means of the flow conditions,
considered in the Reynolds and Weber number, as well as the fuel specifics captured in the Ohnesorge number [26]. Prior to engine simulations, a calibration
of the numerous additional model specific parameters should be conducted.

4.4 Droplet Evaporation

During the droplet evaporation, which is competing with the secondary breakup mechanisms, the liquid fuel continuously changes into its gaseous state, forced by ambient temperature and pressure conditions. Each pressure is assigned a specific temperature at which the fuel evaporates. These are the saturation or vapor pressure p_s and boiling temperature T_b respectively. Changing pressure conditions result in a variation of the boiling temperature. For pure substances, this relation is well characterized by means of the saturation pressure curve. However, liquid fuel like gasoline consists of hundreds of components with varying boiling temperatures, mainly depending on their number of carbon atoms. The molecular interaction of these components characterizes the fuel specific boiling curve, which describes the evaporation rate in dependence of the temperature at a constant pressure level.[26, 72]

The process of evaporation is strongly influenced by the fuel properties and particularly encouraged by high temperatures, a high partial pressure gradient as well as high relative velocities between the droplet and the surrounding gas (enhancing the convective mass and heat exchange [26]). This is, in turn, positively affected by the injection pressure as well as the prevailing flow field (e.g. high swirl or tumble motion). Moreover, break up mechanisms cause smaller droplet sizes which in turn increase the specific fuel surface and therefore its evaporation rate.

Exchange processes between the droplet and the surrounding gas occur on the basis of conservation laws, represented by the equations for mass, momentum and energy in Chapter 2.3. STAR-CD provides a multiphase flow modeling framework, which also includes models to describe the interphase transfer [15]. During the evaporation, mass transfer, which causes a decrease of the droplet size, is accompanied by heat transfer due to the differences in temperature. Here, thermal energy is extracted from the surrounding gas, resulting in a reduction of the ambient temperature ("spray cooling") [71].

QuickSim moreover provides an evaporation model that takes into account the effect of flash boiling. This is a sudden droplet burst and immediate evaporation caused by an abrupt decrease of the ambient pressure below the temperature dependent saturation pressure of the fuel component. This occurs for example if heated fuel exits the injector into a low pressure regime and can have significant influence on the characteristic spray pattern and propagation as presented in [69, 81].

Depending on the injection timing, different influencing factors on the fuel evaporation are prevailing. A direct injection of fuel during the intake stroke is characterized by high relative velocities due to the strong charge motion (high tumble level, turbulence, etc.). During the compression stroke, however, the in-cylinder pressure and temperature level increases, whereby the temperature rise represents the dominant influencing factor [26].

5 Utilized Engine Models

The fuel injection analyses presented in this work were carried out on different project-specific engine models, which will be introduced in the following. Although some units have been modified during the tests, only the standard configurations are listed. Deviations from this will be stated separately, if relevant. Corresponding boundary conditions, in the form of pressure and temperature specifications, were applied on the particular mesh endings, either taken from 1D-CFD simulations or test bench data. The latter were also used to calibrate the injector and engine models in preparation for the actual investigations.

5.1 Injection Chamber

The vast majority of injection analyses were conducted in a virtual injection chamber. Its standard 3D-CFD mesh is displayed in Figure 5.1 (left) and comprises a constant volume of approximately 3.5 liter. The computational domain is discretized into 113,248 hexahedral cells, having an incremental discretization length. The cross section plot of Figure 5.1 (right) indicates that the mesh is refined towards the injector area, having a minimum edge-length of 1.5 mm. The injector coordinate system is centrally positioned in the upper chamber wall, whereby its orientation and mounting depth can be optionally varied. More informations about the injection chamber can be taken from Table 5.1.

The spray analysis within an injection chamber allows an evaluation of the fuel spray propagation under steady state conditions and without any influence of the charge motion or limiting combustion chamber walls. In particular, influencing parameters on macroscopic spray properties, i.e. axial and radial penetration, spray cone angle and single jet orientation, can be thoroughly investigated.

© Springer Fachmedien Wiesbaden GmbH, part of Springer Nature 2019
M. Wentsch, *Analysis of Injection Processes in an Innovative 3D-CFD Tool for the Simulation of Internal Combustion Engines*, Wissenschaftliche Reihe Fahrzeugtechnik Universität Stuttgart, https://doi.org/10.1007/978-3-658-22167-6_5

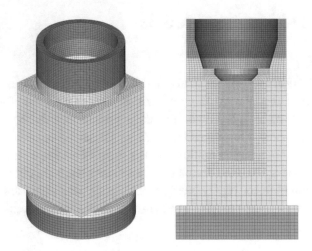

Figure 5.1: 3D-CFD mesh of the injection chamber

Ideally, corresponding optical measurements serve as reference for an injector calibration in preparation of engine simulations. The virtual chamber setup can therefor be adapted very well to varying pressure and temperature conditions. Result files can be written for any desired time step, enabling an optical evaluation of the fuel spray propagation in accordance to the available test bench images.

Table 5.1: Standard simulation configuration of the injection chamber

Number of cells	113,248
Cell discretization length l_c	min. 1.5 mm (injector area)
Time step Δt	12.5 μs
CPU-time	approx. 15 min for 10 ms
Chamber pressure p_{ch}	1 bar (absolute)
Chamber temperature T_{ch}	293 K

5.2 Turbocharged 4-Cylinder DISI-Engine

This turbocharged 4-cylinder spark-ignition engine with direct injection by Volkswagen Motorsport GmbH was developed for the World Rally Championship (WRC). As presented in [56, 75, 79] this development was essentially supported by 3D-CFD simulations. A multitude of analyses enabled a specification of the engine concept, including channel geometry, valve timing, piston shape, injection strategy, etc. Its specifications in accordance to the regulations of the Fédération Internationale de l'Automobile (FIA) (appendix J, article 255A-2014, Art. 5.1 [21]) are listed in Table 5.2. However, due to confidentiality reasons, not the entirety of information can be provided at this point.

Table 5.2: WRC engine specifications

Displacement	1597 cc
Bore	83 mm
Stroke	73.8 mm
max. compression ratio	12.5:1
Valves per cylinder	4
Firing order	1-3-4-2
Max. boost pressure	2.5 bar
Max. engine speed	8500 min^{-1}

The computational mesh shown in Figure 5.2, represents one development variant of the engine. The mesh endings were positioned as far from the cylinders as possible. The intake boundary conditions are defined downstream the air restrictor just before the throttle position, the exhaust ones are located at the turbine and wastegate inlet. In between are approximately 380,000 hexahedral computational cells, whereby 120,000 are included in the cylinders. More information on the engine simulation can be taken from Table 5.3.

Table 5.3: Standard simulation configuration of the WRC engine

Number of cells	380,000 (31,000 in cylinder)
Av. cell discretization length l_c	2.8 mm (1.9 mm in cylinder)
Time step Δt	0.5 °CA
CPU-time (1 operating cycle)	6 - 8 hours
Convergence	After 3-5 cycles
Operating point	6000 rpm, WOT
Type of injection	Direct injection
Injection pressure	200 bar
Type of fuel	Gasoline

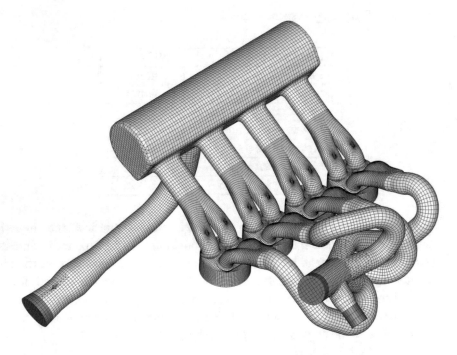

Figure 5.2: 3D-CFD mesh of a turbocharged 4-cylinder DISI-engine

5.3 Single-Cylinder DISI-Engine

Corresponding to the 4-cylinder race engine in Chapter 5.2, the single-cylinder research engine was set up at the LVK of Technische Universität München in collaboration with Volkswagen Motorsport GmbH. Its configuration is, just as the full engine one, in accordance to the FIA WRC regulations. The main specifications are summarized in Table 5.4.

Table 5.4: Single-cylinder research engine specifications

Displacement	400 cc
Bore	83 mm
Stroke	73.8 mm
Max. compression ratio	12.5:1
Valves per cylinder	4
Max. boost pressure	2.5 bar
Max. engine speed	$8500\ \text{min}^{-1}$

The utilization of a single-cylinder test bench, as a pre-development tool or additional investigation device, offers some major advantages over the operation of a full engine setup. Its main field of application can bee seen in the combustion process development. This includes in particular the calibration of (turbo)charged combustion processes with direct fuel injection, since the thermodynamic conditions of the full engine can be captured very well here [51]. The influence of different geometry variants, e.g. of channels, pistons and cam profiles, control and injection strategies can be thoroughly tested. New engine parts only need to be manufactured for one cylinder unit, which makes the operation of a single-cylinder device by tendency less expensive. Due to better accessibility, additional measuring equipment can be integrated easily, e.g. endoscopic measuring equipment for a visual analysis of fuel injection and combustion. [39, 75, 79]

The reliability or rather transferability of single-cylinder analyses is possibly limited for naturally aspirated engines or port fuel injection. Both concepts are highly regulated by the gas dynamics which may deviate from that of the full engine, due to missing cylinders as well as a disparate airbox geometry. More details on the comparability between single-cylinder and full engine devices as well as their 3D-CFD simulation can be taken from [75, 79].

The primary computational mesh of the single-cylinder research engine comprised a flow domain starting from the throttle up to the junction of both exhaust channels. The mesh, as displayed in Figure 5.3, is composed of approximately 123,000 hexahedral cells. This represents approx. one third of the corresponding full engine mesh size (see Section 5.2), having an average cell edge length of 3.9 mm (1.8 mm inside the cylinder).

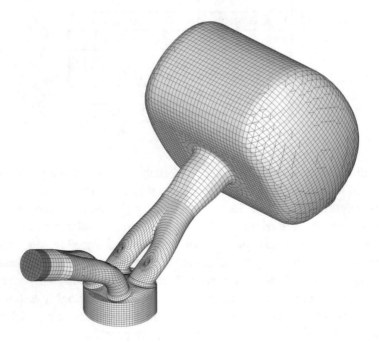

Figure 5.3: 3D-CFD mesh of a single-cylinder DISI-engine

A first validation with available measurement data showed strong deviating pressure traces inside the intake system as well as persistent cycle-to-cycle variations in the air consumption [79]. In order to achieve better conformity, an intake-side extension of the domain seemed reasonable. According to [16], this enables the formation of a convergent flow field and therefore allows a reliable and reproducible simulation of the engine. Consequently, the modeled flow domain has been extended by a settling tank, as illustrated in Figure 5.4, following the actual test bench setup. Here, the intake boundary conditions can be defined at a less critical position, which has a more uniform pressure profile over time.

The extended mesh, illustrated in Figure 5.4 contains approximately one and a half times the cells. However, the computational time did not significantly increase and takes no longer than 3 to 4 hours for one engine operating cycle. The result evaluation showed an immediate improvement of the pressure regime, emphasizing the necessity of utilizing the extended domain for the 3D-CFD simulations. Corresponding information is summarized in Table 5.5.

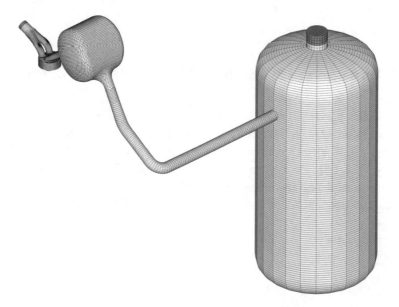

Figure 5.4: Extended 3D-CFD mesh of a single-cylinder DISI-engine

Table 5.5: Standard simulation configuration of the single-cylinder DISI-
engine

Number of cells	185,000
Av. cell discretization length l_c	9.9 mm (1.8 mm in cylinder)
Time step Δt	0.5 °CA
CPU-time (1 operating cycle)	3-4 hours
Convergence	After 3 cycles
Operating point	6000 rpm, WOT
Type of injection	Direct injection
Injection pressure	200 bar
Type of fuel	Gasoline

5.4 Turbocharged 2-Cylinder DI Weber MPE 850 DOHC

The Weber engine employed for the CNG investigations is a 2-cylinder SI-
engine, installed at the Robert Bosch GmbH (department "Gasoline Systems").
It was utilized for comprehensive experimental studies on the direct injection
of CNG [62, 63]. The engine specifications are listed in Table 5.6. In order to
provide more details on the fuel behavior, particularly in scavenging operation
mode, additional 3D-CFD simulations were carried out [64, 76].

Table 5.6: Weber engine specifications

Displacement	846 cc
Bore	89 mm
Stroke	68 mm
Compression ratio	10.5:1
Valves per cylinder	4
Peak cylinder pressure	100 bar
Injection system	DI, centrally mounted
Charging system	Turbo-charged, el. wastegate
Valve train	Intake/exhaust VVT

The computational grid of the 2-cylinder engine, shown in Figure 5.5, includes both cylinders, the intake-system starting upstream the throttle, and the exhaust-system ending upstream the turbine.

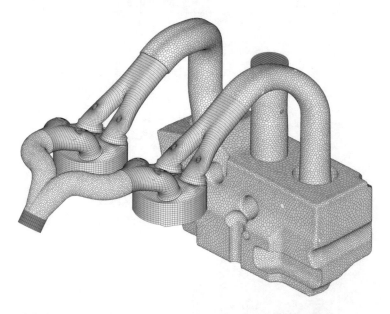

Figure 5.5: Extended 3D-CFD mesh of a 2-cylinder DISI-engine

The grid consists of approximately 300,000 hexahedral cells, whereby only 35,000 are located in each combustion chamber. The standard simulation configuration is summarized in Table 5.7.

Table 5.7: Standard simulation configuration of the Weber engine

Number of cells	300,000
Av. cell discretization length l_c	2.5 mm (cylinder: 1.9 mm)
Time step Δt	0.5 °CA
CPU-time (1 operating cycle)	4 hours
Convergence	After 3 cycles
Operating point	2000 rpm, BMEP 12 bar
Valve overlap	30°
Type of injection	Direct injection
Injection pressure	70 bar
Type of fuel	Methane

6 Numerical Boundary Conditions

A comprehensive analysis of purely numerical influencing factors on the fuel injection forms the basis of this work. The majority of conducted simulations utilized variants of the virtual injection chamber, introduced in Chapter 5.1. This enables an isolated examination of the injection process and ensures a reliable reproduction of conditions for the purpose of comparability. Main spray evaluation criteria in the present study are macroscopic spray properties, which are schematically illustrated in Figure 6.1. These include

- Spray cone angle ε
- Axial spray tip penetration P_a
- Radial spray penetration P_r
- Vortex structures
- Single jet orientation (in case of a multi-hole injector)

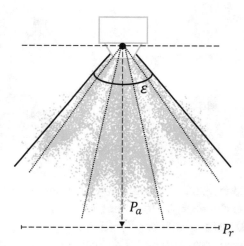

Figure 6.1: Macroscopic fuel spray properties

© Springer Fachmedien Wiesbaden GmbH, part of Springer Nature 2019
M. Wentsch, *Analysis of Injection Processes in an Innovative 3D-CFD Tool for the Simulation of Internal Combustion Engines*, Wissenschaftliche Reihe Fahrzeugtechnik Universität Stuttgart, https://doi.org/10.1007/978-3-658-22167-6_6

For the analyses presented in the following chapters, two different injector types were used: an outward-opening (hollowcone) [55] for liquid and gaseous injection and five symmetric 5-hole injectors for the injection of gasoline, all designed and manufactured by Robert Bosch GmbH. The results of their respective calibration (including one 5-hole injector) by means of optical measurements are shown in Figure 6.2 and Figure 6.3.

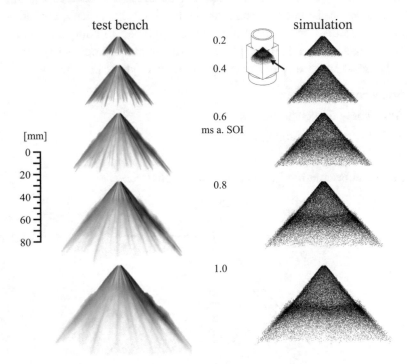

Figure 6.2: Calibration result of gasoline injection into a constant volume
injection chamber (hollowcone injector; $l_{c,\mathrm{min}}$=1.5 mm,
Δt=1/80 ms; p_{inj}=200 bar, T_f=293 K, p_{ch}=1 bar, T_{ch}=293 K; optical images provided by LVK, Technische Universität München)

These represent the starting point of the parameter variations discussed in this chapter, i.e. a reference axial and radial penetration is determined and then transferred to the other parameter variants. The calibration parameters include the spray angles (which, as described in Chapter 8, tend to deviate from the targeting values), droplet initial velocity and size, etc. A single-component gasoline (compare Chapter 7 for more details) for the liquid injection and pure methane for the gaseous injection were implemented as substitute fuel models.

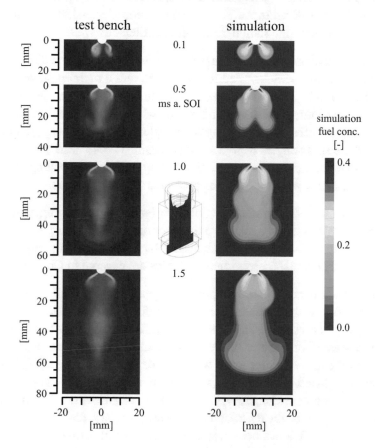

Figure 6.3: Calibration result of methane injection into a constant volume injection chamber (hollowcone injector; l_c=0.75 mm, Δt=1/80 ms; p_{inj}=110 bar, T_f=293 K, p_{ch}=1 bar, T_{ch}=298 K; LIF-images provided by Robert Bosch GmbH)

The Mie-shadowgraphy images (averaging of 5 shots) show distinct irregularities in the form of a stringy breakup of the spray cone. This can originate either from deposits, considerable abrasion or even from the manufacturing process itself and results in an irregular spray penetration. These peculiarities did not have to be considered in the 3D-CFD simulation. Instead, the averaged spray propagation over time should be captured sufficiently well.

In order to achieve conformity, in particular for the axial and radial penetration, the drag model coefficient had to be adapted and a representative number of parcels had to be injected (compare [35]). In addition, it was necessary to implement an additional distribution function which determines the initial droplet velocities. A variation of the spray cone angles over time was not required here, instead the spray angles were defined constant: $\alpha = 43°$ and $\gamma = 3°$. [32] and [81], however, indicate that a variation during the injector opening phase needs to be considered at lower injection pressures (approx. 50 bar).

The calibration of a gaseous injection is by far more challenging, since the fuel propagation reacts more sensitive to the choice of injection parameters. The consideration of varying spray angles in the beginning of the needle lift, for example, is essential (corresponding studies are covered in Chapter 8.2).

At this point it needs to be emphasized again that the modeling of gaseous fuel injection with QuickSim does not aim for a highly detailed replica of supersonic shock waves as in [3, 41, 52, 58]. The reasons for this have already been discussed in Chapter 4.2.2. In order to ensure a macroscopic conformity with the spray images, even without calculating the injector internal flow or using high-resolution meshes, the implementation of a Langrangian droplet approach (compare Chapter 4.2.2) is therefore indispensable. An alternative approach would be the definition of an inlet boundary, directly initializing the gaseous fuel. This has been tested initially but failed due to two main reasons: the lack of reasonable boundary conditions and the inability of capturing the actual spray penetration. Even with the Lagrangian approach, the penetration could only be achieved by assuming the fuel jet to be fully expanded and defining a correspondingly low density as initial value.

6.1 Numerical Stability and Convergence

6.1.1 Injector Position

The first simulations of hollowcone gasoline injection revealed an unexpected discontinuity in the spray evolvement. As illustrated in Figure 6.4 (top left), the spray shows blank gaps along the symmetry axes of the injection chamber (indicated by arrows), by which the cone structure seems to be divided into four. Neither a change in the injection conditions nor a modification of geometric parameters had any impact on the observed phenomenon. Only a relocation of the injector coordinate system into the cell center did significantly improve the spray structure as shown in Figure 6.4 (top right).

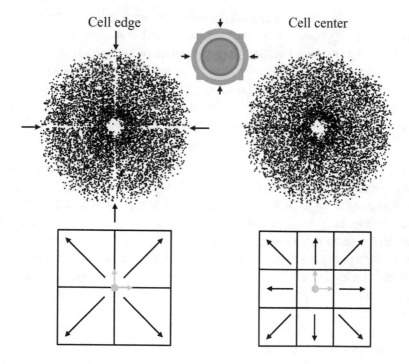

Figure 6.4: Dependence of spray propagation on injector position (0.3 ms a. SOI; schematic illustration following [32])

An appropriate explanation could be found in [32]. The fuel droplet motion is influenced by the flow direction and velocity in the surrounding cells. Averaged velocity vectors are illustrated in two simplified two-dimensional plots in Figure 6.4 (bottom). Defining the injector coordinate system on cell edges, results in a targeted droplet motion towards the four quadrants, whereby a fluid propagation along the symmetrical axes is being prevented. The cell center position, on the contrary, ensures an evenly homogenized flow field, which interacts with the liquid droplets, and prevents the formation of gaps in the spray pattern.

6.1.2 Droplet Initialization Domain

The idea of a droplet initialization domain, first introduced in Chapter 4.2.1, was not only implemented to increase numerical stability. It better represents the injector orifice position (in dependence of the coordinate system position), additionally considers the area of prevailing primary droplet breakup and, which is of utmost importance, compensates the discontinuous injection process due to discrete time steps in the simulation. In order to understand the sensitivity of a fuel spray to a modification of the droplet initialization domain, a variation of L_{max} was simulated, keeping $L_{min} = 0$ mm constant.

Figure 6.5 (images increased to 220 % of calibration reference) shows an exemplary selection of L_{max}-variants, indicating the strongly deviant behavior of the fuel spray at 0.1 ms a. SOI, whereby $L_{max} = 3$ mm is the calibrated reference case. In this early stage of the injection, L_{max} does not only have an influence on the spray penetration (larger L_{max} results in a deeper penetration) but also on the spray pattern for small L_{max}, which can be explained by the initial droplet velocity derived from the injection pressure. Using eq. 4.2 (Bernoulli, energy conservation), a maximum flow velocity through the injector orifice of approx. 232 m/s results from an injection pressure p_{inj} of 200 bar ($p_{ch} = 1$ bar). Further assuming a frictionless flow, the fuel droplets progress about 2.9 mm per time step ($\Delta t = 0.0125$ ms). Even considering the loss of momentum, the choice of $L_{max} = 1$ mm causes a disc-like initialization of droplets due to the high droplet velocity.

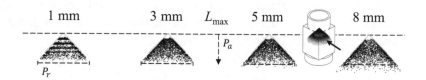

Figure 6.5: Variation of the droplet initialization domain (hollowcone injector; 0.1 ms a. SOI; $l_{c,min}$=1.5 mm, Δt=1/80 ms; p_{inj}=200 bar, T_f=293 K, p_{ch}=1 bar, T_{ch}=293 K; reference: L_{max} = 3 mm)

Despite the different penetration depth at 0.1 ms a. SOI, a uniform penetration can be observed at 0.5 ms (Figure 6.6) and 1.0 ms a. SOI (Figure 6.7, images decreased to 75 % of calibration reference). The influence of L_{max} can merely be observed in the injector vicinity of L_{max} = 1 mm. However, the strong deviations right after the SOI are well compensated, resulting in an equal spray propagation for 1,3 and 5 mm. In contrast to the initial trend, L_{max} = 8 mm shows approx. 5 % less penetration in axial and radial direction.

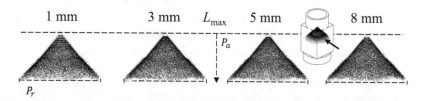

Figure 6.6: Variation of the droplet initialization domain (hollowcone injector; 0.5 ms a. SOI; $l_{c,min}$=1.5 mm, Δt=1/80 ms; p_{inj}=200 bar, T_f=293 K, p_{ch}=1 bar, T_{ch}=293 K; reference: L_{max} = 3 mm)

In addition to the optical analysis, the fuel evaporation behavior has been evaluated as well. Here, no significant discrepancies could be determined.

An analysis of a variation of L_{min} provided no additional information since the occurring effects are similar. Increasing L_{min} reduces the initialization domain and this results again in a disc-shaped initialization. An increase of L_{min} and

L_{max} with $\Delta L = 3$ mm corresponds to a relocation of the injector coordinate system since the droplet initialization does not consider a loss of momentum in dependence of the droplets initial position.

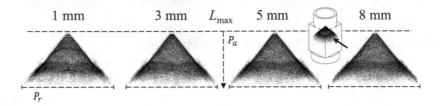

Figure 6.7: Variation of the droplet initialization domain (hollowcone injector; 1.0 ms a. SOI; $l_{c,min}$=1.5 mm, Δt=1/80 ms; p_{inj}=200 bar, T_f=293 K, p_{ch}=1 bar, T_{ch}=293 K; reference: L_{max} = 3 mm)

Although the different L_{max}-variants do converge over time, deviations shortly after SOI can already have an impact on the mixture formation inside the combustion chamber and therefore on the combustion process. Consequently, L_{max} needs to preliminarily be determined in dependence of the injection pressure and the injector specifications. For high pressures and therefore high discharge velocities of the fuel, a time dependent definition of L_{max} further improves the simulation of the fuel propagation in particular at its early phase.

For this work, injections were predominantly investigated at an injection pressure of 200 bar, which led to a consistent choice of $L_{min} = 0$ mm, $L_{max} = 3$ mm.

6.1.3 Lagrangian Droplet Approach

For the purpose of CPU-time saving, the concept of parcels is introduced. This, however, requires a definition of the number of droplets per parcel N_D, which together with the fuel density and specified SMD (15 μm) consequently influences the total number of injected parcels. [35] emphasizes the necessity of a statistically representative number of parcels to reach the spray penetration determined by optical measurements. This can be confirmed by the summarized spray images in Figure 6.8 (images decreased to 55 % of calibration reference), which present the spray propagation over time in dependence of the predefined N_D. For the calibrated reference spray in Figure 6.2, 500 droplets per parcel were defined.

A high N_D (> 2000), and therefore small number of parcels, thins out the spray significantly and results in an underestimation of the spray penetration. Also the small vortex structure on the outer edge of the spray can not be identified anymore. An increase of droplets per parcel moreover slows down the evaporation process, as a comparison of the two extreme cases in Figure 6.9 shows. This can be attributed to the worse statistical representation of the fuel spray by a small number of parcels. The intermediate cases accordingly range in between but are not displayed here due to reasons of clarity.

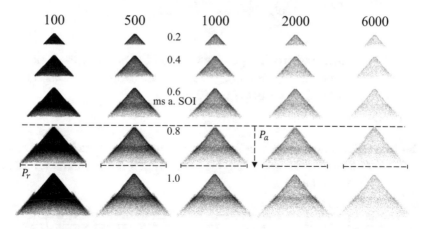

Figure 6.8: Influence of droplets per parcel definition on spray propagation (hollowcone injector; $l_{c,min}$=1.5 mm, Δt=1/80 ms; p_{inj}=200 bar, T_f=293 K, p_{ch}=1 bar, T_{ch}=293 K; reference: N_D=500)

A decrease of N_D (< 500), on the contrary, generates a highly crowded spray. Here, no advantage can be found, which justifies the resulting CPU-time. Table 6.1 provides an overview of CPU-times (simulation of 1 ms) in dependence of the number of parcels. In addition to the CPU-time information, also the total run time is listed, which moreover includes the time for memory allocation, etc. It becomes apparent that the number of droplets per parcel N_D and the CPU-time do not correlate linearly.

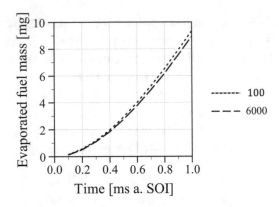

Figure 6.9: Influence of droplets per parcel definition on spray evaporation (hollowcone injector; $l_{c,min}$=1.5 mm, Δt=1/80 ms; p_{inj}=200 bar, T_f=293 K, p_{ch}=1 bar, T_{ch}=293 K)

Table 6.1: CPU-time comparison for different droplet per parcel definitions (1.0 ms a. SOI; m_B=32.95 mg; processor: Intel® Xeon® E3-1245 v3 @ 3.4 GHz)

Droplets/parcel	Number of parcels	CPU-time	Run time
50	541,085	30 min 19 s	33 min 02 s
100	270,535	19 min 48 s	22 min 05 s
500	53,763	9 min 44 s	11 min 37 s
1000	26,929	8 min 40 s	10 min 35 s
2000	13,225	7 min 59 s	9 min 48 s
4000	6,515	7 min 44 s	9 min 37 s
6000	4,257	7 min 36 s	9 min 28 s

Based on the CPU-time for $N_D = 500$, a decrease of N_D (increase of the number of parcels) by factor 10 results in a CPU-time three times as high. However, an increase by factor 10 only reduces the time effort by approx. 22 %. This means a significant reduction of the number of parcels does not provide a worth time benefit but leads to a loss of information, whereas an extreme increase of parcels leads to an unacceptably high time expenditure without revealing any additional information content. This is particularly the case in full engine simulations. Here, also the numerical stability can be negatively affected by a high number of parcels due to the spatial distribution of evaporating droplets, which leads to a higher number of necessary calculation iterations and an increase of the time effort. Nevertheless, experience shows that the parcel demand is usually higher for the simulation of a hollowcone than for a multi-hole injector, since the latter usually has a smaller opening cross section and therefore a lower mass flow rate. A choice of 500 to 1000 droplets per parcels is very convenient for the test case presented in this work.

6.2 Spatial Discretization

As extensively described in Chapter 3, the main characteristic of QuickSim is the CPU-time reduction due to comparatively coarse cell structures. However, a comprehensive analysis regarding the influence on the fuel injection has not yet been carried out. For this purpose, the injection chamber geometry, presented in Chapter 5.1, has been modified. The incremental refinement towards the injector has been replaced by a uniform cell size distribution with a cell edge length l_c of 1.5 mm. This is similar to a cylinder computational mesh in a full engine simulation with QuickSim, which contains manually generated hexahedral cells with preferably uniform cell sizes.

The analysis includes three degrees of spatial refinement, i.e. the reference mesh has been additionally refined in vertical direction (0.75 mm, 0.375 mm). All variants were tested with both, liquid and gaseous fuel. However, neither case involved a supplementary modification of the time increment in order to keep the Courant number constant (more information in the next section).

6.2.1 Liquid Fuel

Figure 6.10 presents the analysis results for the injection of gasoline with a hollowcone injector. No differences in the spray distribution can be observed. The axial and radial penetration, spray cone angle as well as weak vortex structures coincide very well, as exemplified for 0.8 ms a. SOI. Examining the evaporated gas phase progression presented in Figure 6.11 a deviant spray progression can be recognized.

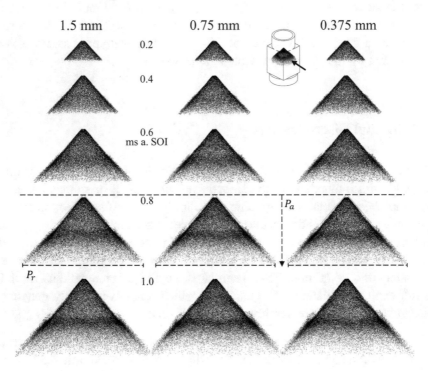

Figure 6.10: Influence of spatial cell discretization on liquid fuel propagation (hollowcone injector; $\Delta t=1/80$ ms; $p_{inj}=200$ bar, $T_f=293$ K, $p_{ch}=1$ bar, $T_{ch}=293$ K; reference: $l_c=1.5$ mm)

A decrease of the cell size seems to favor an increased volumetric spread of gaseous fuel inside the spray cone. The "coarse" mesh has a higher local concentration of fuel instead. This deviation can not be justified by the evaporated fuel mass. With a maximum variance of 1.8 % (at 1.0 ms a.SOI) and by tendency higher evaporation rates for $l_c = 0.75$ mm these are marginal and can be neglected.

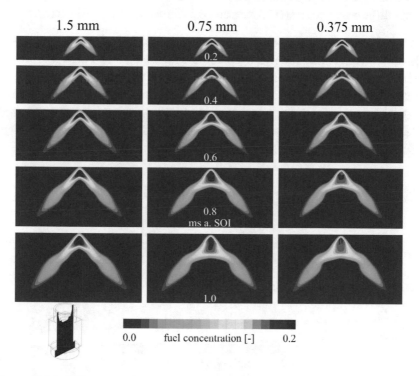

Figure 6.11: Influence of spatial cell discretization on evaporated fuel propagation (hollowcone injector; Δt=1/80 ms; p_{inj}=200 bar, T_f=293 K, p_{ch}=1 bar, T_{ch}=293 K; reference: l_c=1.5 mm)

A section through the hollowcone spray reveals the liquid fuel distribution within the spray cone. Presenting the section plots at 0.8 ms after SOI, Figure 6.12 indicates that the refined meshes (especially $l_c = 0.75$ mm) have more droplets deflected from the main direction of propagation, which results in the higher gaseous fuel concentration in the inner spray area. Larger cells seem to decelerate the background fluid, whereby the air entrainment momentum overcomes the one induced by the injected fuel. An axial progression of the fuel, forced by the resulting flow conditions, is consequently prevented. In the refined cases, the air entrainment is countered by the higher impulsive flow initiated by the fuel injection, resulting in a horizontal flow boundary. A significant difference between 0.75 mm and 0.375 mm can not be detected.

The analyses were additionally repeated with a higher chamber temperature T_{ch} of 400 K, favoring the fuel evaporation. Figure 6.13 and 6.15 indicate that the trend observed before does not change despite the by 60 % increased evaporation, which again does not significantly change for the different levels of spatial discretization (< 1 % at 1.0 ms a. SOI). Again, the central accumulation of evaporated fuel enhances from 1.5 mm to 0.75 mm but does not notably change from 0.75 mm to 0.375 mm.

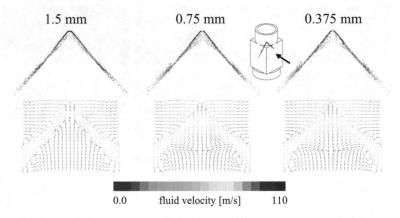

Figure 6.12: Influence of spatial cell discretization on liquid fuel propagation (hollowcone injector; 0.8 ms a. SOI; Δt=1/80 ms; p_{inj}=200 bar, T_f=293 K, p_{ch}=1 bar, T_{ch}=293 K; reference: l_c=1.5 mm)

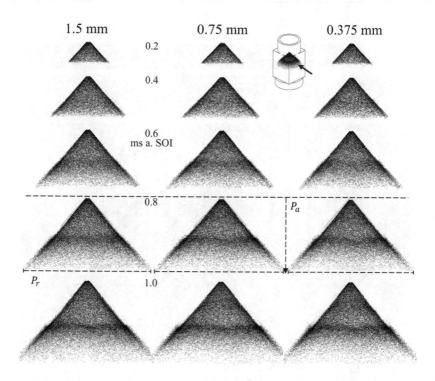

Figure 6.13: Influence of spatial cell discretization on liquid fuel propagation (hollowcone injector; $\Delta t=1/80$ ms; $p_{inj}=200$ bar, $T_f=293$ K, $p_{ch}=1$ bar, $T_{ch}=400$ K; reference: $l_c=1.5$ mm)

Figure 6.14: Influence of spatial cell discretization on liquid fuel propagation (hollowcone injector; 0.8 ms a. SOI; $\Delta t=1/80$ ms; $p_{inj}=200$ bar, $T_f=293$ K, $p_{ch}=1$ bar, $T_{ch}=400$ K; reference: $l_c=1.5$ mm)

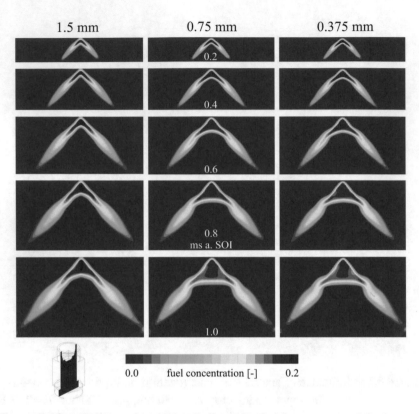

Figure 6.15: Influence of spatial cell discretization on evaporated fuel
propagation (hollowcone injector; Δt=1/80 ms; p_{inj}=200 bar,
T_f=293 K, p_{ch}=1 bar, T_{ch}=400 K; reference: l_c=1.5 mm)

To exclude dependencies on the injector type, a 5-hole injector (which will be
used for studies in Chapter 7) was additionally simulated under standard temperature and pressure (STP) conditions. The evaluated results are summarized
in Figure 6.16. For a multi-hole injector, the individual spray jet orientation
can also be analyzed. Just as the penetration and spray cone angle, the jet
orientation does not show any noteworthy deviations in dependence of the cell
edge length. This applies to the liquid injected fuel as well as to the evaporated
gas phase, which is illustrated by means of fuel concentration section plots in

Figure 6.17. The differences with respect to the evaporation are even lower than for the hollowcone injector (< 0.5 %) and can thus also be neglected.

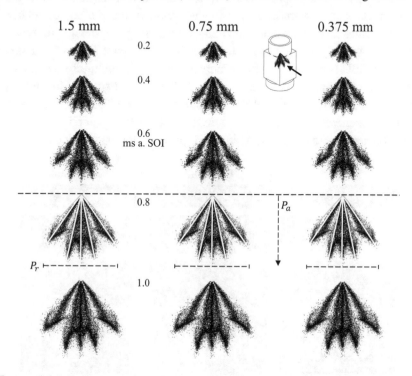

Figure 6.16: Influence of spatial cell discretization on liquid fuel propagation (symmetric 5-hole injector; Δt=1/80 ms; p_{inj}=200 bar, T_f=293 K, p_{ch}=1 bar, T_{ch}=293 K; reference: l_c=1.5 mm)

In summary, the spatial discretization of a 3D-CFD mesh does merely influence the evaporated gas phase. The liquid fuel droplets injected into a constant volume chamber, however, are not affected. A reason for this can be found in the Lagrangian formulation of the liquid fuel injection implemented in the CFD code. Here, the droplet motion follows mesh-independent trajectories, which result from the droplet initial conditions as well as the mutual interference with the surrounding fluid and its induced flow field [15]. Thus, in determining the cell discretization for a full engine simulation, priority must not be given to the liquid fuel injection. Rather, it must be ensured that the ambient (turbulent) flow field in which the droplets move is calculated accurately. For the pure purpose of a macroscopic injector calibration, the use of relatively coarse meshes is reasonable, especially considering the drastic increase of CPU-time and in particular of the total run time (compare Tables 6.2 and 6.3). While the CPU-time shows a linear increase with the total number of cells, the run time for the finest discretization ($l_c = 0.375$) rises disproportionately. Under additional consideration of the Courant number and thus modification of the time increment, the time effort increases even further, which makes such a level of discretization impracticable for engine development purposes.

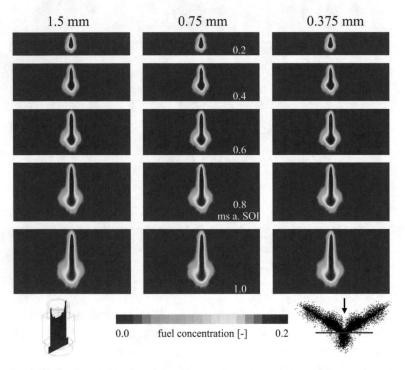

Figure 6.17: Influence of spatial cell discretization on liquid fuel evaporation (symmetric 5-hole injector; $\Delta t = 1/80$ ms; $p_{inj} = 200$ bar, $T_f = 293$ K, $p_{ch} = 1$ bar, $T_{ch} = 293$ K; reference: $l_c = 1.5$ mm)

Table 6.2: CPU-time comparison for different degrees of calculation mesh refinement (hollowcone injector; $\Delta t=1/80$ ms; $p_{inj}=200$ bar, $T_f=293$ K, $p_{ch}=1$ bar, $T_{ch}=293$ K; processor: Intel® Core® i7-2600 @ 3.4 GHz)

Cell discretization	Number of cells	$\Delta t=1/80$ ms	1 ms
Incremental	112,672	11.23 s	0 h 14 min 58 s
$l_c = 1.5$ mm	679,616	47.12 s	1 h 02 min 49 s
$l_c = 0.75$ mm	1,319,616	92.89 s	2 h 03 min 51 s
$l_c = 0.375$ mm	2,599,616	186.65 s	4 h 08 min 52 s

Table 6.3: Run time comparison for different degrees of calculation mesh refinement (hollowcone injector; $\Delta t=1/80$ ms; $p_{inj}=200$ bar, $T_f=293$ K, $p_{ch}=1$ bar, $T_{ch}=293$ K; processor: Intel® Core® i7-2600 @ 3.4 GHz)

Cell discretization	Number of cells	$\Delta t=1/80$ ms	1 ms
Incremental	112,672	11.5 s	0 h 15 min 20 s
$l_c = 1.5$ mm	679,616	49.08 s	1 h 05 min 26 s
$l_c = 0.75$ mm	1,319,616	96.65 s	2 h 08 min 52 s
$l_c = 0.375$ mm	2,599,616	695.15 s	15 h 26 min 08 s

6.2.2 Gaseous Fuel

Occurring differences in the evaporated fuel propagation make it even more reasonable to also run the simulations with gaseous fuel (methane). The results presented in Figure 6.18 point out very clearly that the cell discretization has a distinct influence on the spray behavior. This is an unequivocal contrast to the results of liquid fuel injection.

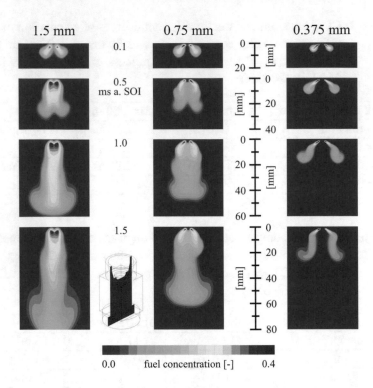

Figure 6.18: Influence of spatial cell discretization on gaseous fuel propagation (hollowcone injector; $\Delta t=1/80$ ms; $p_{\mathrm{inj}}=110$ bar, $T_f=293$ K, $p_{\mathrm{ch}}=1$ bar, $T_{\mathrm{ch}}=298$ K; reference: $l_c=0.75$ mm)

Based on the calibrated reference with an edge length l_c of 0.75 mm, the respective spray parameters (particularly the spray angles α and γ) have been adopted unmodified for simulations with one refined ($l_c = 0.375$ mm) and one coarsened ($l_c = 1.5$ mm) computational mesh.

The different variants already differ shortly after SOI. A mesh refinement effects a more precise initialization within the pre-defined domain (compare Chapter 4.2). The further progression happens primarily in a radial direction, whereby the fine cell structure seems to decelerate the fuel. On the contrary, the initially widened shape within the coarse mesh instantaneously leads to a collapse of the spray, which in turn results in a narrow spray beam and deeper penetration of the fuel. In order to explain this highly divergent behavior, a schematic depiction of the discretization impact on the spray is illustrated in Figure 6.19. Unlike liquid fuel, the gaseous fuel motion is bound to the mesh cells. The fuel concentration information is associated to the computational point P (introduced in Chapter 2.3) and therefore to the entire cell. The conservation equations for mass, momentum and energy are solved for each cell, defining the gaseous fuel transportation.

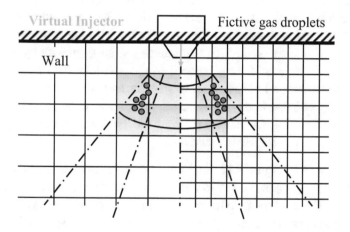

Figure 6.19: Schematic illustration of the spatial discretization influence on gaseous fuel initialization and propagation

After initializing the fictive fuel droplets within the pre-defined domain, these evaporate immediately within the first iteration. The cell in which the droplet was located then contains the respective mass of gaseous fuel. In Figure 6.19, the cells on the right side of the scheme have half the edge length of the cells on the left side. Although the defined droplet initialization domain is identical, the cumulated volume (or surface area for the 2D image) of the cells containing droplets and therefore the gaseous fuel is nearly half for the smaller cells.

For coarse cell structures, the definition of the spray angles as well as L_{min} and L_{max} appear void and the 3D-CFD mesh can not reflect the geometrical specifications. Therefore, the cell discretization needs to at least consider the dimension of the injector orifice, or rather the spray expansion along the injector opening cross section, as characteristic length. But again, a suitable compromise between mesh refinement (along with temporal refinement) and computing time has to be found, based on the required resolution. Besides, an additional refinement may cause numerical instabilities and does not necessarily enhance the results quality, which is then even more influenced by a suitable choice of model parameters. These should at some point be derived from an additional (coupled or decoupled) simulation of the injector internal flow in order to avoid further calibration effort. Such a simulation, however, can contain a mesh consisting of more cells for only one partition of the injector [9, 51] than are needed for a full engine simulation with QuickSim and moreover needs to consider cavitational effects, modeled by a two-phase flow. This computational effort is not acceptable for engine development purposes. Furthermore, utilizing highly refined meshes in favor of the gaseous fuel injection process can limit the validity of other models implemented in QuickSim, i.e. for the wall heat transfer calculation.

The comparison in Figure 6.18 emphasizes that a sole change of the discretization degree is no universal key to a successful injection simulation. It can not compensate for a thorough calibration of the model parameters such that the macroscopic spray structure is well reproduced. Although calculations with QuickSim are preferably run with comparatively coarse meshes, they can still apply a minimum necessary level of refinement in order to enable representative and predictive simulations of gaseous fuel injection in a full engine.

6.3 Temporal Discretization

The calculation time increment can not be chosen arbitrarily. A choice must rather be made on the basis of the flow field and the convective velocities v prevailing therein. As addressed before, the Courant number C provides a stability criterion for convergent CFD simulations.

$$C = \frac{\Delta t \cdot v}{l_c} < 1 \qquad \text{eq. 6.1}$$

The time step Δt is therefore limited and needs to be reduced with decreasing cell size l_c [16, 29]. However, for engine applications this criterion can hardly be satisfied for all cells within the computational mesh [41].

6.3.1 Liquid Fuel

For the present study, the variance of temporal discretization, maintaining the cell size, was limited due to convergence problems. This affected in particular a time increment of 1/20 ms as well as 1/10 ms. Their initial spray development is compared to the reference increment of 1/80 ms in Figure 6.20.

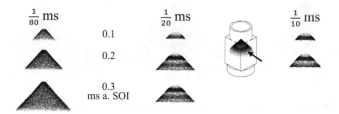

Figure 6.20: Influence of temporal discretization on liquid fuel propagation (hollowcone injector; $l_{c,min}$=1.5 mm; p_{inj}=200 bar, T_f=293 K, p_{ch}=1 bar, T_{ch}=293 K; reference: Δt=1/80 ms)

Since such a discretization is not practical for engine applications, the very distinct influence on the spray propagation shall not be discussed in detail. However, similar, even if alleviated, effects can be observed for a time step $\Delta t = 1/40$ ms (compare Figure 6.21).

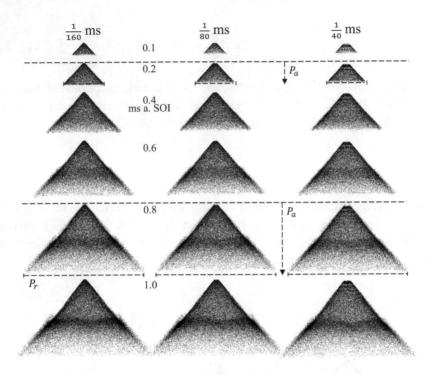

Figure 6.21: Influence of temporal discretization on liquid fuel propagation
(hollowcone injector; $l_{c,min}$=1.5 mm; p_{inj}=200 bar, T_f=293 K,
p_{ch}=1 bar, T_{ch}=293 K; reference: Δt=1/80 ms)

Here too, the droplet velocity overtakes the time increment which results in a
rudimentary disc-shaped initialization and therefore a loss of the upper hollowcone structure. In addition, for the time step $\Delta t = 1/160$ ms, a deeper
penetration is apparent at timings close to SOI. An explanation for this can
be found in the definition of a droplet initialization domain by means of L_{min}
and L_{max}. These do not change with the temporal refinement. For smaller time
steps, fewer droplets are initialized into the calculation mesh more frequent.
Due to their velocity and the negligence of a momentum loss in dependence
of the initial droplet position, the droplets exceed the distance of L_{max} from
the injector position. At a time, where droplets would be first initialized for
a bigger time step, these would still be located within the limits of L_{min} and

L_{max}. An additional graphical evaluation at 0.8 ms a. SOI, however, indicates that these initial differences get compensated and have therefore no negative influence on the further liquid spray evolution.

Examining the evaporated gas phase in Figure 6.22 reveals equally marginal distinctions. Solely the smallest time increment shows a higher fuel concentration in the spray center. The diagram in Figure 6.22 additionally quantifies the evaporated fuel mass over time and confirms this observation. With decreasing time increments, the fuel tends to evaporate faster, which results in up to 5.4 % (3.1 %) more gaseous fuel mass for $\Delta t = 1/160$ ms in comparison to $\Delta t = 1/40$ ms ($\Delta t = 1/80$ ms) at 1.0 ms a. SOI. This deviant behavior, however, can not be traced back to the evaporation modeling since the evaporated fuel mass is calculated independent from the chosen time step Δt.

To test whether the phenomenon observed in the injection chamber simulation also has influence on the fuel evaporation and mixture distribution in an engine operating cycle simulation, the single-cylinder DISI-engine (Chapter 5.3) was simulated with two different Δt. Here, the time increment of 0.5 °CA ($\hat{=}$ 0.0139 ms at 6000 rpm) represents the standard choice for QuickSim full engine simulations.

Exemplary evaluations presented in Figure 6.23 show the evaporated fuel mass in dependence of time. In accordance to the injection chamber analysis, the gasoline evaporates by tendency faster for the smaller time increment (0.125 °CA) during compression phase. At ignition point (IP), 0.75 mg ($\hat{=}$ 1.2 %) more fuel is evaporated. In spite of the significantly higher temperatures, the relative deviation of gaseous fuel mass is smaller compared to the injection chamber investigations.

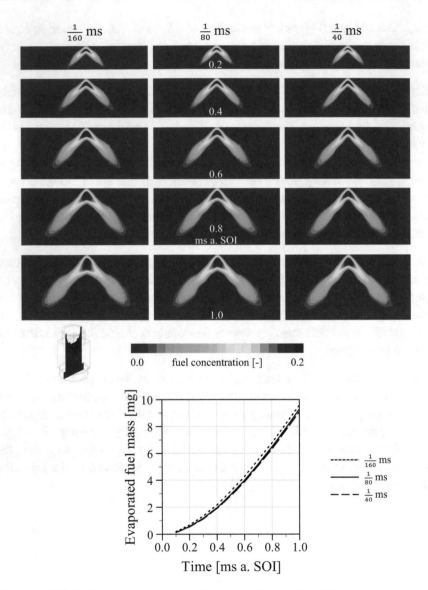

Figure 6.22: Influence of temporal discretization on liquid fuel evaporation (hollowcone injector; $l_{c,min}$=1.5 mm; p_{inj}=200 bar, T_f=293 K, p_{ch}=1 bar, T_{ch}=293 K; reference: Δt=1/80 ms)

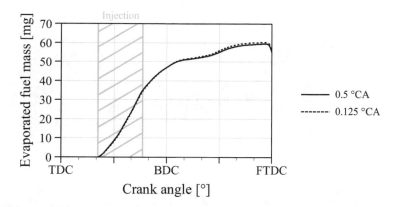

Figure 6.23: Influence of temporal discretization on liquid fuel evaporation
(Single-cylinder DISI-engine; 6000 rpm, WOT; p_{inj}=200 bar,
T_f=350 K)

Optical analyses of the mixture formation are illustrated in Figure 6.24. A
comparison of the cylinder lambda at IP shows that the fuel is mostly well
distributed in both cases. Differences only occur in the area between the intake valves and along the outer valve edges. While a lean area develops on
the intake side for a time step of $\Delta t = 0.125$ °CA, the simulation of a larger
time increment ($\Delta t = 0.5$ °CA) shows an additional rich region, indicating fuel
accumulation in the squish area. This, however, only has minor influence on
the global engine cycle parameters as summarized in Table 6.4, which presents
the changes caused by a time step refinement from 0.5 to 0.125 °CA from one
cycle to the next. With an IMEP variation of 0.5 %, these lie within acceptable
cycle-to-cycle variations and show that the deviating evaporation process plays
a subordinated role within the engine simulation. Nevertheless, it is explicitly
recommended to simulate different engine variants only with the same temporal discretization in order to guarantee comparability. Since [59] moreover
emphasizes that the time step has a significant influence on the combustion
process, it should ideally be kept constant over the entire engine operating
cycle.

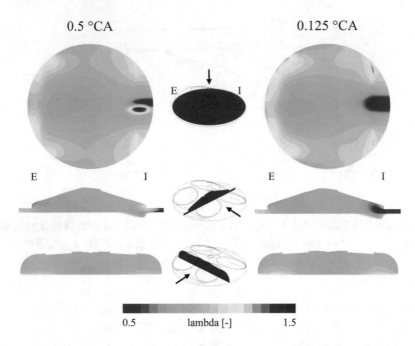

0.5 °CA   0.125 °CA

0.5   lambda [-]   1.5

Figure 6.24: In-cylinder lambda distribution at IP for different time increments (Single-cylinder DISI-engine; 6000 rpm, WOT;
p_{inj}=200 bar, T_f=350 K)

Table 6.4: Engine cycle comparison for different time steps Δt (0.125 °CA in
comparison to 0.5 °CA; single-cylinder DISI-engine; 6000 rpm,
WOT; p_{inj}=200 bar, T_f=350 K)

Air mass m_A (IP)	+ 0.2 %
Fuel mass m_B (IP)	+ 1.2 %
λ (IP)	- 0.93 %
5-95 %	+ 1.5 %
p_{max}	- 1.4 %
IMEP	+ 0.54 %

6.3.2 Gaseous Fuel

Since gaseous fuel simulations do require a finer calculation mesh, only smaller time increments were compared with the reference $\Delta t = 1/80$ ms. In fact, Δt larger than $1/80$ ms did cause distinct convergence problems with a cell edge length of 0.75 mm and could therefore not be analyzed.

According to the images in Figure 6.25 a temporal refinement at first effects a less expanded but further progressed fuel jet (compare t = 0.1 a. SOI). This causes the spray cone to collapse later, which finally leads to a decrease in the penetration depth.

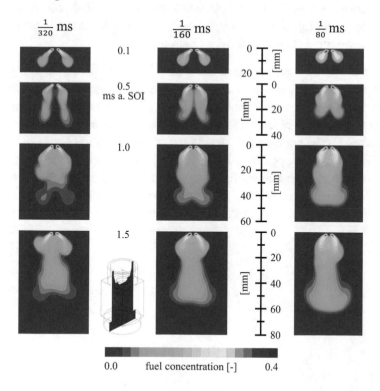

Figure 6.25: Influence of temporal discretization on gaseous fuel propagation (hollowcone injector; l_c=0.75 mm; p_{inj}=110 bar, T_f=293 K, p_{ch}=1 bar, T_{ch}=298 K; reference: Δt=1/80 ms)

This behavior is similar, though less distinct, to the cell size reduction. The explanation for this, however, is different. In both cases, the deviant penetration can be put down to the fuel initialization. In case of spatial refinement, finer meshes can more precisely reproduce the geometrical specifications (which, due to the interaction with other input parameters, does not necessarily result in a better spray calibration). Smaller time increments, however, are sensible to the fuel initial domain. For the implemented Lagrangian description of fictive gas droplets in QuickSim, an initialization domain is defined in the same way as for liquid fuel injection. This domain, again, does not change along with the mesh refinement. Therefrom resulting deeper penetration in the beginning of the injection, as described above, initiates the modified fuel propagation. For the gaseous fuel injection, an adaptation of the initialization region needs to come along with a change of the time increment.

In order to ensure a reliable and stable simulation, in particular for gaseous fuel, the time step should be chosen in numerical accordance to the cell discretization. On the basis of these conditions, the calibration of the remaining parameters can take place. If this does not lead to a satisfactory conformity (depending on the application), the choice of discretization needs to be adapted. This procedure applies to liquid as well as gaseous fuel injection. Full engine simulations should finally be set up in consideration of the entirety of calibrated boundary conditions. Conformity with other models implemented in QuickSim has thereby always to be fulfilled.

7 Liquid Fuel Modeling

The modeling of highly complex in-cylinder processes is characterized by the employment of severe simplifications. This is done, in particular, in the numerical description of fuel. The fact that liquid fuel consists of a multitude of components, for example, is commonly neglected in most multi-dimensional models. Instead, single-component surrogate fuels are employed.

The implementation of fuel models in QuickSim calls for similar simplifications to ensure conformity with the tool's characteristics. However, in order to prevent a loss of information and to capture essential fuel properties, depending on the subject of investigation, diverse modeling approaches are deployed.

The following descriptions are limited to the chemical-physical fuel properties broached in Chapter 4.1.1 for the characterization of spray propagation, atomization and evaporation of liquid fuel. Properties, which have an immediate influence on the fuel combustion process (e.g. calorics, laminar flame speed, etc.) were not subject of the investigations.

7.1 Single-Component Fuel Modeling

7.1.1 Literature Approaches

For engine applications, which exceed pure research purposes, it is still common practice to employ fuel models consisting of a single component. This can either be a synthetic model in accordance with the specific fuel properties or a pure chemical substance. Depending on the application, a standard component can either represent the chemical or physical characteristics. Gasoline is therefore usually substituted by iso-octane or n-heptane respectively [24, 41]. In Table 7.1, their properties are listed in comparison to a racing gasoline (RON = 100.3), which was used for most of the analyses presented in this work.

© Springer Fachmedien Wiesbaden GmbH, part of Springer Nature 2019
M. Wentsch, *Analysis of Injection Processes in an Innovative 3D-CFD Tool for the Simulation of Internal Combustion Engines*, Wissenschaftliche Reihe Fahrzeugtechnik Universität Stuttgart, https://doi.org/10.1007/978-3-658-22167-6_7

Table 7.1: Thermo-physical properties of racing gasoline (RON = 100.3),
n-heptane and iso-octane (values of n-heptane and iso-octane originate from [73])

Properties	T [°C]	Gasoline	N-heptane	Iso-octane
ρ [kg/m^3]	15	747.0	690.91	690.0
η [Pa·s]	25	$4.12 \cdot 10^{-4}$	$4.04 \cdot 10^{-4}$	$5 \cdot 10^{-4}$ (20°C)
σ [N/m]	25	0.0279	0.0199	0.01877 (20°)
c_p [J/kg·K]	25	2,227.12	2,238.28	2,122.82
h_v [J/kg]	25	275,996.46	365,655.76	269,543.9
p_s [Pa]	37.8	59,167.29	11,133.55	5,865.00 (40 °C)
T_b [°C]	-	37.0 - 147.0	98.4	99.2

Commercial CFD-codes usually provide databases which include synthetic
or standard single-component models, ideally considering their temperature
dependency. Alternatively, the properties can be user defined by means of
subroutines. For pure substances, the empirically derived formulas in [73]
(compare Appendix A2) enable a sufficiently accurate modeling. Moreover,
thermodynamic properties of chemical substances can be taken from relevant
literature like [50] or various online or commercial databases, e.g. [19, 43, 57],
whereby not all of them include calculation formulas or temperature varying
information.

7.1.2 Synthetic Gasoline Model

Gasoline properties can not be considered consistent. In fact, they vary depending on the crude materials, manufacturing process, additives, etc. A universal replacement by n-heptane or iso-octane is therefore not ideal. Table 7.1,
for example, indicates that some significant differences exist in comparison to
the racing gasoline. Particularly noteworthy is the deviating saturation pressure, which influences the evaporation process. In order to better match the
fuel specifics, a single-component synthetic surrogate was modeled, whereby
a comprehensive fuel analysis of the racing gasoline served as reference. For
each thermo-physical property (compare Chapter 4.1.1), a pure substance was

chosen which showed the best conformity with the analyzed data. In this way, the empirical formulas in [73] could be utilized to make the temperature dependent data available for the 3D-CFD simulation. Only for the saturation pressure an alternative procedure was followed.

- **Density** ρ [kg/m^3]
  The fuel density measurement can theoretically span a temperature interval from -20 °C up to 120 °C. An increase in the saturation pressure, however, can cause bulk boiling due to cavitational effects, which can disturb the measurements and limits its significance. As indicated in Figure 7.1, the implementation of n-undecane (paraffin, $C_{11}H_{24}$) results in an adequate conformity of the few density values.

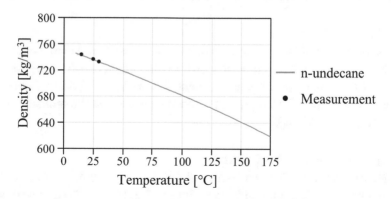

Figure 7.1: Approximation of fuel density with n-undecane

- **Dynamic Viscosity** η [Pa·s]
  As with the density, the same measurement difficulties can occur for the fuel viscosity. This limits the analyzed interval, which could involve temperatures from 25 °C to 120 °C. Based on the very few measuring points, cyclopentane (naphthene, C_5H_{10}) was chosen as substitute component. A comparison of measured racing fuel values with the viscosity of cyclopentane is given in Figure 7.2.

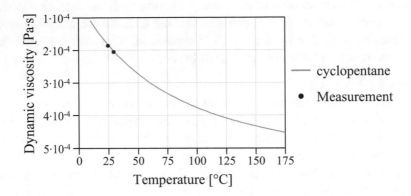

Figure 7.2: Approximation of fuel viscosity with cyclopentane

- **Surface Tension** σ [N/m]
  A measurement of the surface tension is hardly possible, due to the negative influence of the fuel decomposition with rising temperature. Since the variance between individual fuel components is very low, an arbitrary choice of one is reasonable. For the present work, the choice was the aromatic hydrocarbon toluene (C_7H_8).

- **Specific Heat Capacity** c_p [J/kg·K]
  Since measurement data was not available for the specific heat capacity, a component was chosen, which according to the data in [73] approximately represents average values of the entirety of fuel components. N-hexen (olefine, C_6H_{12}) is therefore a good representative.

- **Saturation Pressure** p_s [Pa]
  Figure 7.3 shows that a very good approximation of the saturation pressure curve could be achieved by means of a fourth degree polynomial (eq. 7.1). Due to the temperature limit for the measurement method (approximately at 110 °C), this polynomial additionally provides an extrapolation for the missing pressure values at higher temperatures.

$$p_s(T) = 1.43 \cdot 10^{-4} \cdot T^4 - 3.12 \cdot 10^{-2} \cdot T^3 - 28.74 \cdot T^2$$
$$+ 11.56 \cdot 10^3 \cdot T - 11.60 \cdot 10^5$$

eq. 7.1

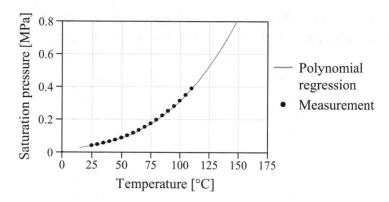

Figure 7.3: Polynomial saturation pressure approximation

- **Heat of Vaporization** h_v [J/kg]
  For pure substances, the heat of vaporization can be determined arithmetically from the saturation pressure curve, using the "Clausius-Clapeyron equation" [50], which describes a temperature-dependent change of the saturation pressure considering the molar evaporation enthalpy. The implementation for mixtures, however, is not necessarily reasonable. In addition, since the pressure of saturation grows linearly over the measured temperature range (correlation coefficient r > 0.9999), the heat of vaporization can be regarded constant for temperatures between 25 °C and 110 °C [ASG Analytik, e-mail from January 12, 2017]. However, in order to provide data for a wider temperature range, n-heptane (paraffin, C_7H_{16}) is regarded as appropriate model substitute for h_v.

7.1.3 Limitations in the Application of Single-Component Fuel Models

Although synthetic single-component surrogates enable a better fuel specific modeling than standard components, they are still limited in their reproduction of mixture properties [41]. In particular, highly temperature and pressure dependent characteristics may not be replicated by a single-component model. As described in Chapter 4.4 liquid fuels do not have a fixed boiling point like pure substances, they rather have a boiling curve which results from the boiling temperatures of the individual hydrocarbon compounds [26]. For the before introduced racing gasoline, the boiling curve is displayed in Figure 7.4.

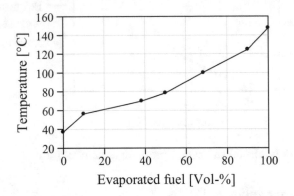

Figure 7.4: Boiling curve of reference racing gasoline

The fuel evaporation progress is accompanied by a constant decomposition of the fuel. Due to the varying component evaporation behavior, a change of the mixture composition takes place namely a gradual enrichment of low volatile components as illustrated in Figure 7.5. This causes an additional variance in the fuel properties over time and clarifies that even single-component models which consider temperature dependency are not sufficiently accurate for certain internal combustion engine applications.

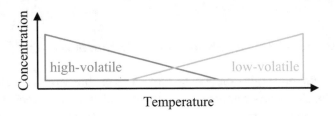

Figure 7.5: Fuel decomposition due to varying component volatility

Injection related phenomena, like evaporation, wall film or even soot formation, may not be simulated realistically with single-component models [24, 41,
53], even if fuel model inaccuracies are tried to be compensated by additional
calibration effort. For some purposes it is consequently mandatory to employ a
multi-component fuel model, consisting of high and low volatile hydrocarbons,
to improve the predictive capabilities. An adequate choice of components is
still able to reduce the complexity of real fuel while better reflecting essential characteristics, especially its spray breakup, evaporation and propagation
behavior.

7.2 Multi-Component Fuel Modeling

For applications which involve transient processes, it is anticipated that effects
related to the character of fuels as component mixtures are not negligible. Here,
multi-component models might be more convenient than single-component
substitutes. In this section, the modeling and implementation of such models will be discussed, followed by an application-based comparison of single-
and multi-component surrogate fuels in Section 7.3.

7.2.1 Implementation of Multi-Component Fuel Models

In general, two different ways of modeling multi-component fuel substitutes
exist. Both methods, the continuous and the discrete modeling, are able to
capture the temporal variability of fuel properties along with the change of
temperature and pressure conditions. The fundamental difference is schematically illustrated in Figure 7.6.

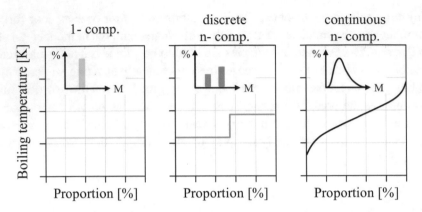

Figure 7.6: Comparison of fuel modeling approaches [24]

The continuous modeling, which was first introduced by [68], employs a probability density function $f(p) \in [0, 1]$ to describe the liquid fuel composition [33, 37] based on a characteristic property p which differs for individual fuel components, e.g. the molecular weight [24]. Further fuel properties like the boiling curve can be deduced therefrom. According to [53], the CPU time demand of a continuous fuel model is comparatively low while maintaining the predictive capability in terms of multi-component evaporation and fuel decomposition. For the simulation of a continuous model, fuel specific parameters have to be empirically determined, which explicitly characterize the fuels' initial composition and evaporation behavior. For more information on the continuous modeling as well as an extension thereof, the reader may refer to [24, 41, 80].

Following the idea of discrete modeling, the liquid fuel is described by a finite number of components whose composition is defined by respective mass fractions. For each thermo-physical property, the mixture characteristics are determined by those of the individual components. As exemplary illustrated in Figure 7.6, the evaporation is rather described by boiling steps than a boiling curve.

A model implementation in QuickSim always follows the demand for versatile applicability as well as model simplicity without sacrificing the required accuracy. For the fuel modeling, this implies an enabling of application specific transitions between single- and multi-component surrogate fuels, since in some cases a mixture-based model is not necessarily needed. The discrete method meets these demands and furthermore allows an indicative simulation of injection related and influenced processes. Here, conclusions concerning every implemented hydrocarbon compound can be drawn [53] and the decomposition in the course of evaporation is not only apparent by a shift in the molecular weight. A utilization of thermo-physical databases additionally simplifies the modeling and modification of implemented fuel models. The CPU demand for this approach increases with the number of components [24]. However, according to [36], most fuels can be modeled sufficiently accurate by less than 10 components.

For the composition of a discrete fuel model, which is capable of reflecting the thermo-physical properties (in particular the fuel evaporation), it is generally recommended to use pseudo-components. These serve as representatives for groups of fuel components with similar characteristic properties [24]. The classification can for example be geared to the molecular weights, boiling temperatures or carbon atoms. As pseudo-components, either synthetic surrogates or pure substances can be implemented. Here, a choice of components from the same chemical group is sufficient for the sole description of the evaporation characteristics [8]. In order to also represent a correct $C_nH_mO_rN_q$ distribution, etc., representatives from diverse chemical species should be included in the model.

Even with detailed knowledge of all fuel compounds, identifying a discrete component composition which coincides with the real fuel properties is no trivial endeavor. [26] emphasizes that the molecular interaction of the components essentially influences their evaporation behavior, i.e. the mixture behaves differently than a simple linear combination of all components and the superposition principle is not applicable. The mutual interference rather causes e.g. a shift of the component evaporation, as illustrated in Figure 7.7 by means of component specific evaporation rates in comparison to their boiling temperatures. Particularly interesting is that low volatile components already start to evaporate at lower temperatures, while the contrary happens with high volatile components, which have by tendency a delay in their initial evaporation.

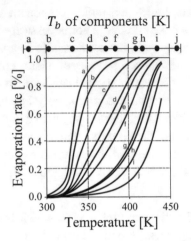

Figure 7.7: Simulated evaporation behavior of a 10-component fuel model
          [26]

7.2.2 Literature Approaches for Gasoline

Although gasoline consists of hundreds of different hydrocarbon molecules,
which depending on their chemical structure can be distinguished by varying
characteristics, it is not expedient to include a majority of these into a multi-
component fuel model. At some point, no additional benefit is created but
the CPU demand still increases. Literature provides various gasoline models,
utilizing between 2 and 7 components, some of which will as an example be
presented in the following.

- **Primary Reference Fuels** (PRF)
  The application specific binary mixture of n-heptane and iso-octane, is more
  commonly used for experimental determination of the fuel research octane
  number. Its numerical implementation is mainly focused on auto-ignition or
  knock prediction simulations as well as on the modeling of reduced kinetic
  reaction mechanisms. For spray investigations, PRF are rarely employed due
  to a poor reproduction of the evaporation behavior and inaccurate H/C-ratio.

- **36 % n-pentane, 46 % iso-octane, 18 % n-undecane** (vol.-%)
  The components and their proportions were chosen by [13] in order to match
  the experimentally determined destillation curve of commercial gasoline. A
  minimum of three components is required to reasonably describe the progress of evaporation. Ideally, a high (n-pentane, C_5H_{12}, $T_b = 36.2$ °C) and
  low volatile (n-undecane, $C_{11}H_{25}$, $T_b = 195.9$ °C) component is complemented by a middle one (iso-octane, C_8H_{18}, $T_b = 99.2$ °C).

- **25 % n-pentane, 9 % n-hexane, 11 % n-heptane, 23 % n-octane, 16 %
  n-nonane, 9 % n-decan, 7 % n-dodecane** (mol.-%)
  In order to model the composition of typical gasoline fuel, [53] introduces
  seven paraffins. The predicted evaporation characteristics are in good agreement with the measured data. Therefore the discrete mixture is sufficient
  to e.g. analyze the performance of multi-component evaporation models,
  although the components do not represent all chemical species that can be
  found in real gasoline. For some internal combustion engine applications,
  this approach may not have enough diversified information content.

- **4 % n-butane, 16 % iso-pentane, 25 % toluene, 3 % n-hexane, 32 %
  iso-octane, 18 % 1,2,3-trimethylbenzene, 2 % tridecane** (wt.-%)
  [8] also employs a seven component model, consisting of paraffins and aro-
  matics. The bulk properties of the model are very close to commercial gasoline.

7.2.3 QuickSim Specific Multi-Component Modeling

The same argument concerning individualized modeling that applies to single-
component models takes effect for multi-component approaches. For the purpose of calculation model validation or very general injection analyses, the
utilization of literature approaches is reasonable. In the case of engine development tasks, which are focused on injection strategy analyses and therefore incorporate specific injectors and fuels, an application dependent model
generation is mandatory. At this point, it should be mentioned that the multi-
component modeling in this work refers exclusively to the liquid fuel phase
and its physical properties. The gaseous fuel and its chemical specifics, in

turn, is modeled by means of a single-component gasoline to describe the fuel ignition and combustion. This means, each of the liquid components evaporates into the same gaseous one.

For the racing gasoline, introduced in Section 7.1, a specifically adapted multi-component surrogate fuel was modeled. Since literature research revealed that a number of 3 to 10 components is reasonable to realistically describe fuel, 5 hydrocarbon compounds were selected. In this way, the main chemical species can be considered, the fuel boiling curve can be adequately reproduced and a good compromise between level of detail and computational effort is achieved. At this point, it should be emphasized again that the modeling of the liquid fuel is particularly oriented towards a reliable reproduction of its evaporation behavior. Following the selection of components (pure substances), their mass fractions were determined iteratively until the composition reached a satisfying conformity with the measured saturation pressure curve as displayed in Figure 7.8.

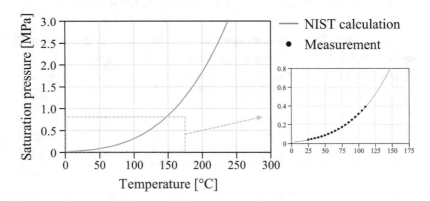

Figure 7.8: Comparison of experimental saturation pressure curve with multi-component model of racing gasoline

In doing so, the mixture saturation pressure curve was calculated by the "NIST Reference Fluid Thermodynamic and Transport Properties Database (REF-PROP)" (version 9.1) [42]. The final choice and composition of the fuel model is listed in Table 7.2. For the 3D-CFD simulation with QuickSim, the thermo-physical properties of each selected hydrocarbon compound are determined by the frequently mentioned equations provided by [73].

Table 7.2: 5-component fuel model for a racing gasoline

Mass fraction	Substance	Chem. formula	Species	T_b [°C]
17 %	iso-pentane	C_5H_{12}	iso-paraffin	27.8
20 %	cyclopentane	C_5H_{10}	naphthene	49.2
36 %	n-heptane	C_7H_{16}	paraffin	98.4
18 %	toluene	C_7H_8	aromatic	110.7
9 %	m-xylol	C_8H_{10}	aromatic	139.1

The vast majority of publications, which make multi-component fuel models a subject of discussion, utilizes them for the description and development of evaporation models for multi-component droplets. These can particularly be distinguished in their level of detail, concerning inner droplet temperature distribution, transportation processes (mass and heat transport), flash boiling regimes, etc. [52]. A consideration of all these processes, in particular additional transport equations as well as equilibrium conditions for each mixture compound, significantly increases the computational effort of discrete multi-component models [53]. [24] specifies the overhead by an increase of CPU-time linearly with the number of components.

Such a demand for computing capacity is not acceptable for a fast response 3D-CFD tool like QuickSim. For this reason, an alternative procedure is implemented. Instead of multi-component droplets, each droplet is only assigned with one fuel compound, similar to the procedure described in [1]. The components are still assumed to form a miscible mixture, where the evaporation (incl. flash boiling) of each droplet occurs in dependence of the other components. As explained in Section 7.2.1, the saturation pressure is a function of the component concentrations in the mixture. Assuming the droplets to be completely unaffected by each other, would result in an incorrect simulation of the fuel evaporation, which in turn limits its predictive capability. A detailed description of the calculation models for evaporation conditions and rate of each component is provided by [15].

The droplet initialization in the case of multiple fuel components is very similar to the procedure described in Chapter 4.2. The only exception affects the parcel and droplet definition, as illustrated in Figure 7.9. To assure uniformity in the initialization procedure, for each parcel (with N_D droplets per parcel) that is initialized per iteration of the single-component injection (or even common multi-component injection), n parcel will be initialized for the n-component model. For each parcel that is initialized for the first component of the model, one parcel is also initialized for each other component. The definition of a component specific number of droplets per parcel $N_{D,i}$, $i \in 1, 2, ..., n$ (weighting of N_D) ensures the correct mass to be initialized per iteration.

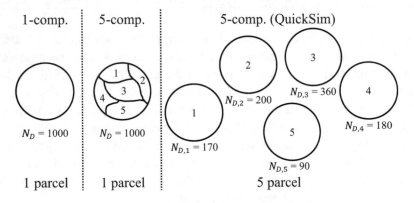

Figure 7.9: QuickSim multi-component droplet initialization

This procedure results in an n-fold increase of the total number of injected parcels for the same fuel mass. In order to prevent a significant increase of the CPU demand, the global quantity of droplets per parcel N_D can be raised, provided that the fuel spray is statistically still well represented. Depending on the resulting number of parcels, it can be reasonable to implement multi-component fuel models only in full engine simulations which have already a converged flow field. This means, the first engine operating cycles will be calculated with a single-component model until the air mass flow does not show strong cycle-to-cycle variations anymore.

7.3 Application Example: Spray Targeting and Evaporation

This analysis aims to determine the influence of different single- and multi-component fuel models on the fuel spray properties. As in Chapter 6, macroscopic spray characteristics are of main interest. Since a temperature and pressure variation is included in the investigations, the fuel evaporation will additionally be determined.

The idea of investigating the fuel spray propagation for different fuel models stems from experimental measurements which were carried out at the LVK of Technische Universität München. Here, diesel fuel and gasoline were injected into a cold constant volume injection chamber under identical conditions and with the same 6-hole fuel injector. Diesel fuel and gasoline differ substantially in their composition, physical properties and thus in spray breakup and evaporation behavior. While gasoline components have an average molecular length of 5 to 12 carbon atoms, diesel fuel consists of more long-chained hydrocarbon molecules with an average length of 10 to 15 carbon atoms. This results i.a. in a higher viscosity and higher boiling temperatures for diesel fuel. Figure 7.10 shows the spray Mie-scattering images for both fuels at different times after the SOI (averaged images of 5 shots). Image differencing (gasoline - diesel fuel) is additionally used, in order to better identify the deviations.

Most noticeable is the different perpendicular spray progression. This can be justified by the higher viscosity and thus later exit of the diesel fuel from the nozzle. The deviations in the 3-dimensional fuel propagation are therefore rather a temporal offset. Apart from that, an unexpected high conformity can be observed in the spray cone angles, which are measured directly at the injector orifice, as well as the single jet directions. This suggests that, under STP conditions, the fuel composition has no immediate influence on the macroscopic spray properties.

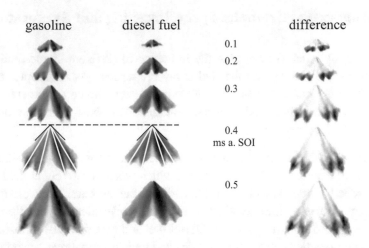

Figure 7.10: Spray propagation comparison of gasoline and diesel fuel using
the same fuel injector (p_{inj}=250 bar, T_f=293 K, p_{ch}=2.5 bar,
T_{ch}=298 K; images provided by LVK)

To verify this assumption by means of 3D-CFD simulations, additional optical
measurements of a 5-hole injector were provided by LVK. These served as
reference for the simulation of various single- and multi-component gasoline
models, given otherwise identical conditions.

The single-component fuel (referred to as 1c gasoline), introduced in Chapter
7.1, was used for an initial injector calibration based on the few geometrical
targeting values provided by the injectors' manufacturer. The calibration results are presented in Figure 7.11. While subsequently maintaining the thereby
determined injection parameters, the fuel model was replaced by additional
single-component substitutes iso-pentane, n-heptane and toluene as well as
before introduced discrete multi-component approaches, which differ in the
number, choice and composition of their components:

- 33.13 % n-pentane, 47.4 % iso-octane, 19.47 % n-undecane (wt.-%) [13]
- 17 % iso-pentane, 20 % cyclopentane, 36 % n-heptane, 18 % toluene, 9 %
  m-xylol (wt.-%, racing gasoline)

- 16.65 % n-pentane, 7.16 % n-hexane, 10.17 % n-heptane, 24.25 % n-octane, 18.94 % n-nonane, 11.82 % n-decan, 11.01 % n-dodecane (wt.-%) [52]

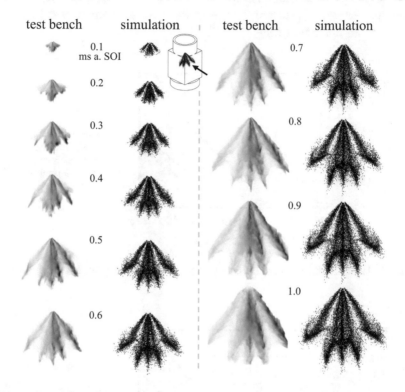

Figure 7.11: Calibration result of gasoline injection into a constant volume injection chamber (5-hole injector; p_{inj}=200 bar, T_f=293 K, p_{ch}=1 bar, T_{ch}=293 K; optical images provided by LVK, averaging of 5 shots)

The evaporation properties of the selected single- and multi-component fuel models are shown in Figure 7.12 and Figure 7.13 respectively. More information on thermo-physical model properties can be taken from Appendix A3. While the curves of single-component substitutes clearly differ from each other, the three-component and five-component models (referred to as 3c and 5c gasoline) correspond in their evaporation despite the different component composition and independent model creation. Also the 1c and 5c gasoline, which both represent models of the initially introduced racing gasoline, coincide as intended in their saturation pressure. However, their diverse composition impact is to be analyzed in the following.

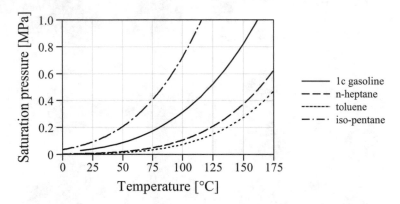

Figure 7.12: Saturation pressure comparison single-component fuel models

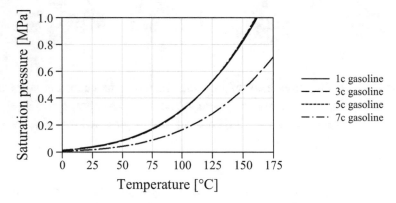

Figure 7.13: Saturation pressure comparison multi-component fuel models

The simulation results of the fuel injection under STP conditions are summarized in Figure 7.14 and Figure 7.15.

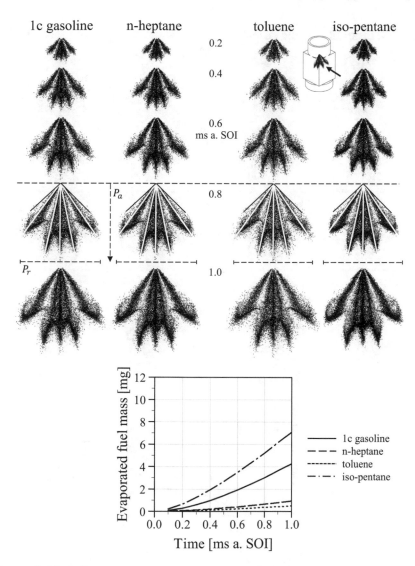

Figure 7.14: Influence of fuel models on spray propagation and evaporation (single-component fuel; 5-hole injector; $l_{c,min}$=1.5 mm,
Δt=1/80 ms; p_{inj}=200 bar, T_f=293 K, p_{ch}=1 bar, T_{ch}=293 K;
reference: 1c gasoline)

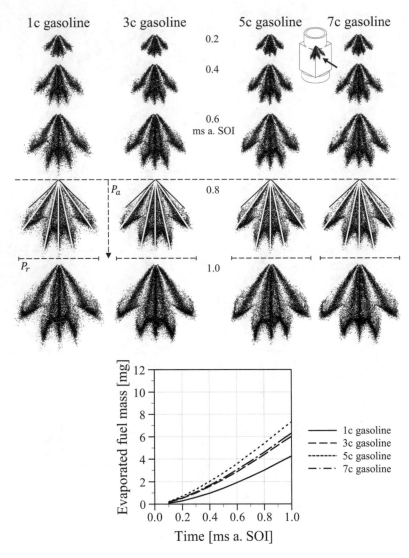

Figure 7.15: Influence of fuel models on spray propagation and evaporation (multi-component fuel; 5-hole injector; $l_{c,min}$=1.5 mm, Δt=1/80 ms; p_{inj}=200 bar, T_f=293 K, p_{ch}=1 bar, T_{ch}=293 K; reference: 1c gasoline)

The graphical comparison indicates that the macroscopic spray propagation is not significantly influenced by the fuel modeling. Optical differences in the compactness of the spray jets can be attributed to the different densities of the individual fuel models and thus varying number of parcels. Spray cone angles, jet directions and penetration values mostly coincide with each other.

A close observation of the single-component models shows that iso-pentane penetrates by tendency lower in axial and radial direction. This can be explained by the rather high evaporation rate of iso-pentane (compare Figure 7.14) which favors a collapse of the individual spray beams. However, their orientation as well as the spray cone angle are not affected by this.

Similar behavior can be observed for the multi-component models in comparison to the 1c gasoline (Figure 7.15). Although their saturation pressure curves were equal or for the 7c gasoline even lower than the one of the reference single-component model, all three models show higher evaporated fuel masses within the analyzed time frame. This can be attributed to their particular composition of individual components. The multi-component models contain high-volatile substances, which evaporate earlier, i.e. at lower temperatures, than the single-component racing gasoline (Appendix A3). As with iso-pentane, this leads to a marginal reduction in the penetration depth. At this point it should be noted that the initialization of the fuel droplets for a multi-hole injector takes place in a random mode. This means the distribution of the droplets does not occur uniformly for all nozzles (just as in reality), which has effects on the spray propagation and symmetry.

[46, 60] confirm that the macroscopic spray properties highly depend on the fuel evaporation and therefore on the diverse boiling temperatures of the individual components. For this reason, the injection simulations were repeated under varying conditions. First, the injection chamber temperature was increased to 373 K in order to favor or even accelerate the evaporation process. Effects on the single- and multi-component models can be examined in Figures 7.16 and 7.17 respectively.

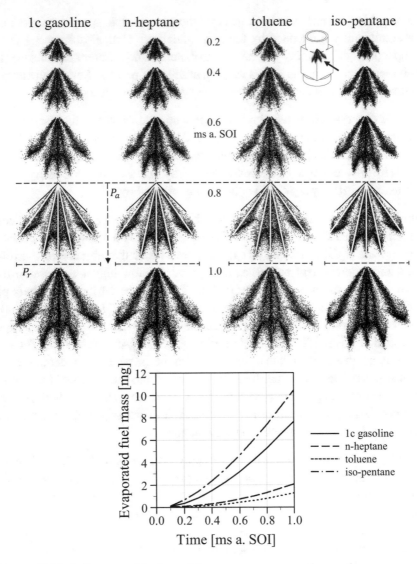

Figure 7.16: Influence of fuel models on spray propagation and evaporation (single-component fuel; 5-hole injector; $l_{c,min}$=1.5 mm,
Δt=1/80 ms; p_{inj}=200 bar, T_f=293 K, p_{ch}=1 bar, T_{ch}=373 K;
reference: 1c gasoline)

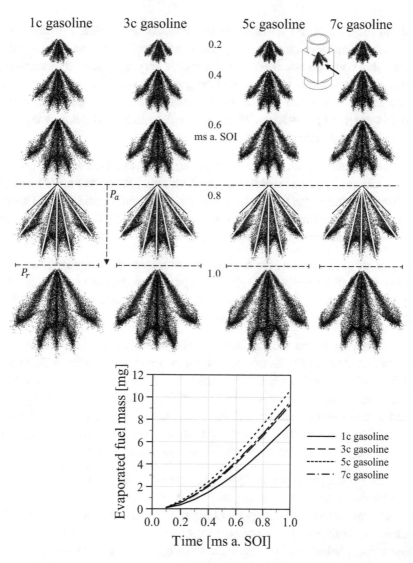

Figure 7.17: Influence of fuel models on spray propagation and evaporation (multi-component fuel; 5-hole injector; $l_{c,min}$=1.5 mm, Δt=1/80 ms; p_{inj}=200 bar, T_f=293 K, p_{ch}=1 bar, T_{ch}=373 K; reference: 1c gasoline)

With an increase of the ambient temperature, the trend does not change for the single-component models. Iso-pentane still penetrates less in comparison to the other components, while these coincide very well in their macroscopic characteristics. However, the 1c and 5c gasoline, which were both modeled on the basis of the same fuel, seem to converge in their progression, particularly in their axial penetration. The other models, in turn, did not change their propagation relative to the 1c gasoline reference.

In spite of the distinct differences between the model properties (see Appendix A3) on the one hand and their component composition on the other hand, the resulting deviations can be considered marginal. Even forcing the fuel models to evaporate more (up to 1/3 of the injected fuel mass), did not significantly change their behavior in relation to each other.

Analysis of the injection into a chamber with increased counter-pressure of 2.5 bar shall additionally determine the spray propagation under boosted SI-engine conditions. Examining Figures 7.18 and 7.19, again, no deviant behavior from the before described can be recognized.

In summary, the purpose of an injector calibration (under any ambient condition) in preparation of ICE simulations can be very well fulfilled with an arbitrary kind of fuel model, whether it consists of one or more components. Since this is usually part of the pre-processing within engine development tasks, it is nevertheless reasonable to try and minimize the undertaken effort. Investigations of the spray targeting should therefore be conducted with the same fuel that is going to be utilized for the following engine applications.

Furthermore, the presented analyses did also point out that even fuel models with a similar saturation pressure profile show a different behavior with respect to their evaporation. This was particularly noticeable for multi-component fuel models, depending on their composition. As described above, for engine development tasks, which are strongly dependent on fuel related processes, special attention has to be paid to the fuel modeling in order to capture essential phenomenon in dependence of the spray breakup, evaporation, deposition, etc.

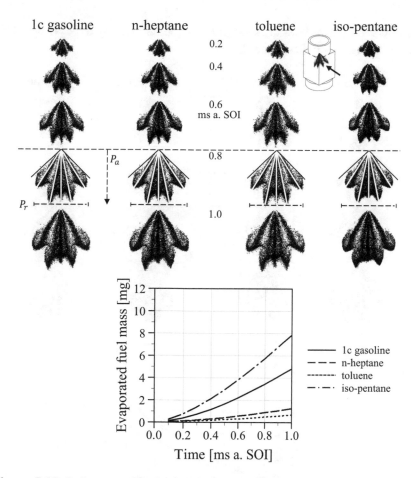

Figure 7.18: Influence of fuel models on spray propagation and evaporation (single-component fuel; 5-hole injector; $l_{c,min}$=1.5 mm, Δt=1/80 ms; p_{inj}=200 bar, T_f=293 K, p_{ch}=2.5 bar, T_{ch}=293 K; reference: 1c gasoline)

In [77], an investigation of wall deposits shows that potential soot formation can be better identified with the use of a multi-component fuel. Due to the overestimated evaporation rate of the single-component model, the wall deposition was by far underestimated, restraining a meaningful comparison of two injector variants both installed in the same turbocharged 4-cylinder DISI-engine, introduced in Chapter 5.2.

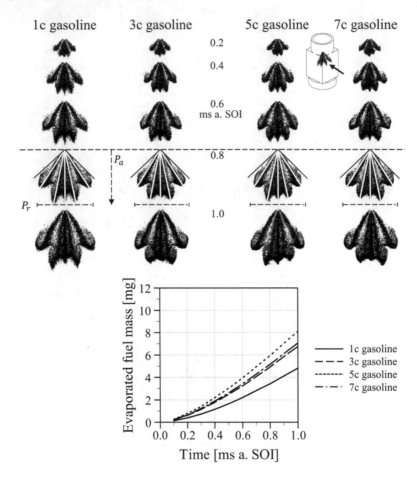

Figure 7.19: Influence of fuel models on spray propagation and evaporation (multi-component fuel; 5-hole injector; $l_{c,min}$=1.5 mm,
Δt=1/80 ms; p_{inj}=200 bar, T_f=293 K, p_{ch}=2.5 bar, T_{ch}=293 K;
reference: 1c gasoline)

The application of a fuel model which contains fractions of low-boiling components, on the contrary, showed results in accordance to the reference particle measurements (Filter Smoke Number) and allowed the analysis of the injector specific spray distribution and wall attachment.

8 Parametrization of Injector Properties

The previous chapters covered numerical boundary conditions as well as fuel modeling as basic influencing factors on the fuel injection simulation. In addition, application or project specific challenges may ensue which need to be mastered. Here, the major challenge is to utilize limited input as best as possible and derive the corresponding injection parameters for the simulation with QuickSim.

[26, 27, 32] and others provide numerous modeling and calculation approaches in order to calculate the initial droplet velocity, droplet size distribution (SMD), time-dependent penetration, etc. The majority of these formulas require injector specific information as input parameters, e.g. cross-section area, flow or discharge coefficient (describing the difference between the theoretical, i.e. calculated, and actual mass flow rate) and others. This data, however, is either not available or can not be directly applied, whereby a thorough calibration of the injector becomes indispensable in order to ensure reliable and in the end predictive injection simulations.

In this chapter, two meaningful examples will be presented, which have been identified during joint projects with Volkswagen Motorsport GmbH and Robert Bosch GmbH. Both involve requirements concerning the parametrization of geometrical injector specifications, for liquid as well as gaseous fuel injection.

8.1 Injector Geometry and Manufacturing

As extensively described in [6, 56, 75, 79], QuickSim has been used to support the virtual engine development process of Volkswagen Motorsport GmbH for years. In order to find an optimal engine configuration in terms of performance, efficiency and drivability, engine geometries (of channels, piston, etc.), valve timings and injection strategies were examined and optimized.

© Springer Fachmedien Wiesbaden GmbH, part of Springer Nature 2019
M. Wentsch, *Analysis of Injection Processes in an Innovative 3D-CFD Tool for the Simulation of Internal Combustion Engines*, Wissenschaftliche Reihe Fahrzeugtechnik Universität Stuttgart, https://doi.org/10.1007/978-3-658-22167-6_8

During an analysis of different injector variants, inconsistencies between the
3D-CFD simulations and corresponding test bench data occurred. Some of the
examined injectors showed mediocre performance in the 3D-CFD calculation
but proved to be efficient at the test bench.

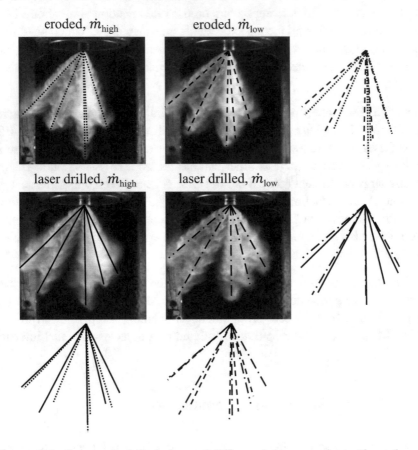

Figure 8.1: Geometrical deviations of different injector variants (frontal
view; 5-hole injector; 0.8 ms a. SOI, $l_{c,min}$=1.5 mm; p_{inj}=200 bar,
T_f=363 K, p_{ch}=1.6 bar, T_{ch}=373 K; optical images provided by
Volkswagen Research Wolfsburg)

The investigated 5-hole gasoline injectors differed in their mass flow rate ("high"/ "low") as well as drilling technology ("spark eroded"/ "laser drilled"). Due to confidentiality reasons, details on the injector geometries can not be provided at this point. Additional injection chamber analyses of the four injector variants revealed a distinct influence on the spray pattern as illustrated in Figures 8.1 and 8.2, which show the frontal and side view respectively.

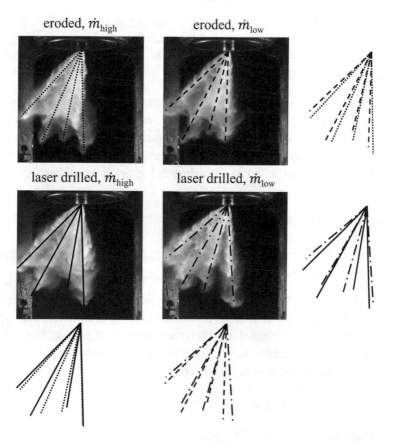

Figure 8.2: Geometrical deviations of different injector variants (side view; 5-hole injector; 0.8 ms a. SOI, $l_{c,min}$=1.5 mm; p_{inj}=200 bar, T_f=363 K, p_{ch}=1.6 bar, T_{ch}=373 K; optical images provided by Volkswagen Research Wolfsburg)

Although the geometrical specifications, provided by the manufacturer, were identical for all variants, the spray cone angles ε and in particular single jet orientations appear very different. An expansion of the fuel spray is effected by both, a higher mass flow rate as well as the laser drilled manufacturing. While the laser drilling affects a 10 % wider jet angle α compared to the specification, the spark eroded one is decreased by approximately 10 %.

These quantifications are specifically determined for the presented study and do not have general validity. However, [30] describes similar observations regarding the influence of laser drilled injector holes, which cause a significant spray widening due to sharper inlet edges.

Due to a lack of reference data, the injectors have initially been simulated in accordance with the geometrical targeting, i.e. spray jet angles and orientation. Figure 8.3 exemplifies a comparison between the test bench images and the simulated fuel spray of the laser drilled injector with high mass flow rate. From both viewing directions, the experimental spray is considerably wider than the numerical one.

Although the simulation input parameters were directly taken from the injector targeting, neither the graphically determined jet axes of the simulated spray nor those of the experimental one, entirely match the targeting, which is reproduced by a CAD image in Figure 8.4.

In order to understand and better reproduce the jet deflection observed at the test bench, additional simulations have been conducted. These included various modifications of the original injection parameters and settings like initial velocity, droplet size distribution, model coefficients influencing the secondary spray breakup and evaporation as well as spray jet angles. None of these could cause the simulated spray to approach to the measured one by deflecting the jet axes from the injector middle axis.

Since the injection modeling with QuickSim does not consider the injector internal flow nor the primary droplet breakup, it was reasonable to assume that phenomena occurring within the injector may be responsible for the different jet orientation, e.g. cavitation ("cold boiling"), fluid detachment and atomization. To verify this, cavitational effects were implemented by a modification of the fuel thermo-physical properties, in particular the heat of vaporization. In addition, a split injection with two co-axial injectors was realized, where part of the gasoline was injected conventionally and the remaining fraction at its vaporization temperature. With this method, various degrees of cavitation as well as changing thermodynamic conditions could be tested. Since again no effect of spray widening could be achieved by this, thermodynamical influences were not found responsible for the jets deflection.

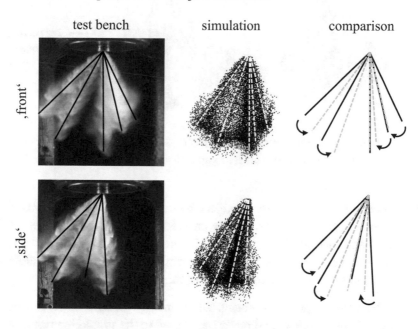

Figure 8.3: Comparison of test bench images to simulated spray with geometrical injector definition (injection chamber; laser drilled 5-hole injector, $\dot{m}_{high}$; 0.8 ms a. SOI, $l_{c,min}$=1.5 mm; p_{inj}=200 bar, T_f=363 K, p_{ch}=1.6 bar, T_{ch}=373 K)

The procedure of exclusion finally attributed the axes deflection to the manufacturing technology. In contrary to thermodynamic phenomena, however, this can not be considered directly within the QuickSim model environment and rather requires an empirical adaption of the geometrical injector settings, including zenith angle θ, azimuth angle φ (see Chapter 4.2) and spray jet angle α if needed.

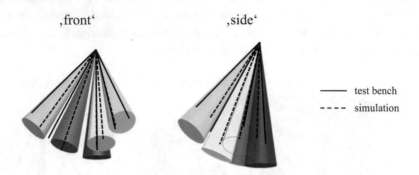

,front'   ,side'

——— test bench
- - - - simulation

Figure 8.4: Experimentally and numerically derived spray jet axes in comparison to the geometrical injector targeting

A modification of the injector geometry by means of the graphically derived jet axes orientation and width, finally provided a satisfying agreement of the simulated and measured spray pattern. Figures 8.5 and 8.6 present the spray propagation over time for both the geometrically defined and the modified injector simulation in comparison to the test bench images.

To demonstrate the good agreement, the delineated jet axes in the test bench image at 0.8 ms a. SOI were directly transferred to the simulation result of the modified spray without further adaption.

The presented analyses showed that injector specifications do not provide sufficient information to ensure a reliable injection simulation and therewith predictive engine calculations. Effects on the spray pattern caused for example by the drilling method can be even more significant then effects of thermodynamic phenomena, which can be included in the QuickSim injection modeling. A thorough calibration of the macroscopic spray characteristics including the jet orientation and width, ideally by means of optical spray images, is therefore even more important.

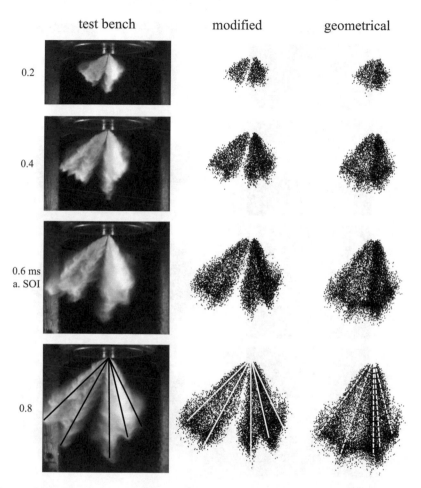

Figure 8.5: Comparison of test bench images to simulated spray with modified and geometrical injector definition (frontal view; laser drilled 5-hole injector, $\dot{m}_{high}$; 0.8 ms a. SOI; $l_{c,min}$=1.5 mm; p_{inj}=200 bar, T_f=363 K, p_{ch}=1.6 bar, T_{ch}=373 K)

In addition to the presented findings, [17] shows how even small adjustments in the injector geometrical layout affects the engine in-cylinder processes. To emphasize the necessity of a calibration as realized above, some of the full engine results are presented in the following.

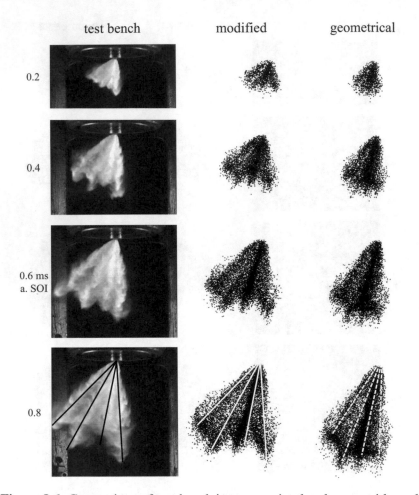

Figure 8.6: Comparison of test bench images to simulated spray with modified and geometrical injector definition (side view; laser drilled 5-hole injector, $\dot{m}_{\text{high}}$; 0.8 ms a. SOI; $l_{c,min}$=1.5 mm; p_{inj}=200 bar, T_f=363 K, p_{ch}=1.6 bar, T_{ch}=373 K)

Here, 3D-CFD simulations of the turbocharged 4-cylinder DISI-engine, introduced in Chapter 5.2, were executed both with the geometrical and modified injector model and compared with a good performing 6-holes spark eroded reference injector. Figure 8.7 shows the fuel distribution for the injectors under investigation at identical operating conditions.

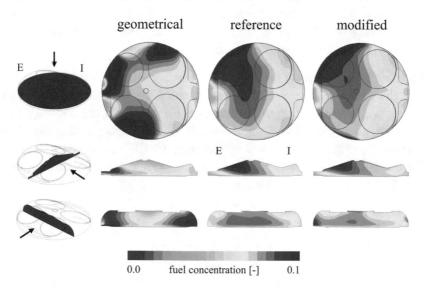

geometrical   reference   modified

0.0   fuel concentration [-]   0.1

Figure 8.7: Comparison of the fuel distribution at IP employing different numerical injector geometries (4-cylinder DISI-Engine, cylinder 1; 6000 rpm, WOT; p_{inj}=200 bar, T_f=363 K)

The geometrical variant, using the targeting geometry provided by the injector manufacturer, shows not only a different but also worse fuel distribution ($\lambda = 1$ is represented by a concentration of approx. 0.07) in comparison to the modified injector which, in turn, coincides well to the reference injector. Although these also have asymmetries in the mixture formation, the spark plug area at least presents a higher fuel concentration. Instead, the geometrical injector causes high fuel concentrations along the liner, particularly on the exhaust side of the cylinder. This results in a noticeable reduction of the simulated engine performance (lower IMEP).

Moreover, a modification of the injector axes orientation seems to significantly support the in-cylinder charge motion. Figure 8.8 compares the tumble level for the different injector variants, whereby not only an improvement of the modified over the geometrical injector can be observed but even in comparison to the reference injector. This can be attributed to the better jet breakup of the (modified) laser drilled injector. An additionally higher level of turbulent kinetic energy, finally justifies the increase in engine power for the modified injector with respect to the reference one.

For more information on the engine analyses, the reader may refer to [17]. With the presented analyses, however, it could be emphasized that diverse influencing factors make a calibration of the injector and the resulting spray propagation indispensable. Only the combination of best possible initial and boundary conditions with a well conditioned 3D-CFD mesh and specifically adapted calculation models enables a reliable and in the end predictive simulation of internal combustion engines with QuickSim.

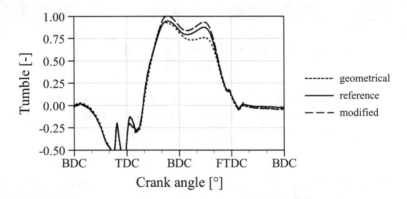

Figure 8.8: Comparison of the tumble level employing different numerical injector geometries (4-cylinder DISI-Engine, cylinder 1; 6000 rpm, WOT; p_{inj}=200 bar, T_f=363 K; non-dimensional)

8.2 Injector Mounting and Opening

Potentials of CNG direct injection to reduce HC emissions were comprehensively analyzed by Robert Bosch GmbH (Gasoline Systems) [62, 63, 64]. Experimental investigations revealed that the gaseous fuel jet is highly sensible to the prevailing conditions. The studies therefore included a variation of boundary conditions (injection pressure, injection timing, needle tip protrusion, etc.) and determined their impact on the jet shape and mixture formation.

An outward-opening hollowcone injector was used for analyses in an injection chamber as well as a 2-cylinder DISI-engine, introduced in Chapter 5.4. For charged operation with scavenging, the jet shape, realized through the injector mounting depth, was identified as one of the dominating factors, influencing the jet propagation, fuel distribution and finally the measured unburned HC emissions.

Figure 8.9 presents engine measurement data for the inward shifted ("+2 mm"), neutral ("±0 mm") and outward shifted ("-2 mm" and "-4 mm") injector positions. The exhaust gas analysis over different SOI shows two peaks in the HC concentration. The first peak can be explained by a too early SOI (360 °CA b. FTDC) during the valve overlap (VOL) phase. Here, the fuel is directly injected into the exhaust channel, independent from the injector position. The second peak, on the contrary, only occurs for the mounting depths of ±0 mm and in particular -2 mm. Additional FFID measurements (fast response flame ionization detector) further revealed that peaks of the HC concentrations inside the intake manifold directly correlate with these HC peaks in the exhaust gas. This leads to the assumption that a back flow of fuel mass into the intake system occurs and the fuel is scavenged into the exhaust channel during valve overlap of the next engine operation cycle.

3D-CFD analysis with QuickSim should help to understand the occurring in-cylinder processes and provide temporal and spatially resolved information about the spray propagation, cylinder charge homogenization, fuel wall attachment and back flow through the intake valves. In preparation of the full engine simulation, it was necessary to calibrate the CNG hollowcone injection. For this purpose, Robert Bosch GmbH provided Schlieren images for the ±0 mm, -2 mm and -4 mm installation positions. +2 mm could not be realized in the injection chamber and was therefore only analyzed in the engine setup.

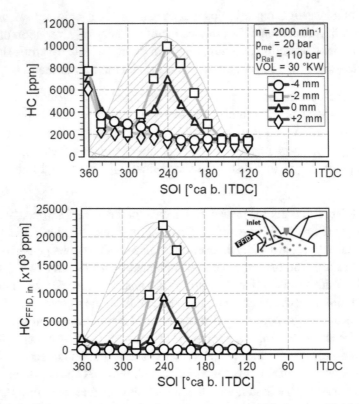

Figure 8.9: HC measurements in the exhaust gas and intake manifold for
different mounting positions and injection timings (2000 rpm,
BMEP=12 bar, λ=1, p_{inj}=110 bar, VOL=30°; data provided by
Robert Bosch GmbH)

Figure 8.10 illustrates the changing fuel propagation in dependence of the
mounting depth. Shifting the injector outward by -2 mm effects strong attachment of nearly the complete fuel mass to the upper chamber walls. Due to
the relative position of the injector opening to the injector bore edge, the fuel
exits close to the chamber surface. This restricts air entrainment which leads
to a flow acceleration and consequently to a local pressure drop (Bernoulli's
law), effecting such a significant jet deflection. Pulling the injector further out
the chamber causes the spray to deflect from the injector bore walls, exiting

therefrom as a fully collapsed beam. In comparison to the standard mounting position, this results in deeper penetration.

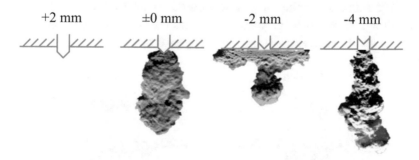

Figure 8.10: Schlieren images of different injector mounting positions (hollowcone injector; CH4 in N2; 0.8 ms a. SOI; p_{inj}=70 bar, T_f=293 K, p_{ch}=1.3 bar, T_{ch}=298 K; optical images provided by Robert Bosch GmbH)

Since the injector is not physically incorporated into the calculation mesh but represented by a coordinate system (as described in Chapter 4.2), the resulting injector depression and protrusion had to be realized via modification of the mesh structure, as illustrated in Figure 8.11.

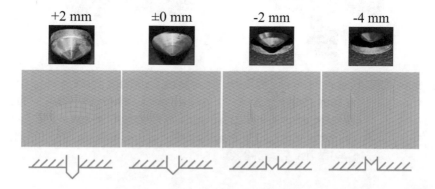

Figure 8.11: 3D-CFD mesh modification for different injector mounting positions inside the virtual injection chamber

The initial calibration attempt was made in accordance with the procedure at that time. The cell discretization was incremental with a minimum edge length l_c of 0.75 mm and the time step Δt was chosen to be 1/80 ms. Apart from the mass flow rate $\dot{m}$, all input parameters were defined constant over time, for example the initial velocity v and spray angles α and γ.

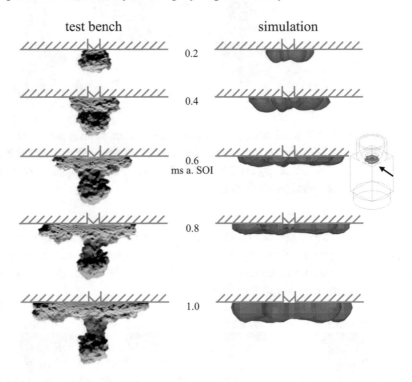

Figure 8.12: Comparison of Schlieren images to simulated spray propagation for mounting depth of -2 mm (hollowcone injector; CH4 in N2; constant spray angle α; iso-surface, threshold value = 5 % fuel concentration; l_c=1.5 mm, Δt=1/80 ms; p_{inj}=70 bar, T_f=293 K, p_{ch}=1.3 bar, T_{ch}=298 K)

Initial simulations of the position variants showed significant deviations compared to the Schlieren images, concerning the axial fuel penetration, the spray shape and its evolution over time. As can be seen in Figure 8.12, the discrepancy is particularly noticeable for the -2 mm position. Here, the fuel propagation is illustrated by means of surface plots with a threshold value of 5 % (i.e. 95 % of the fuel is located within the volume) to better allow a comparison to the Schlieren images. The simulation captures the radial propagation along the upper injection chamber wall. The axially penetrating gas cloud, however, is missing in the calculation results.

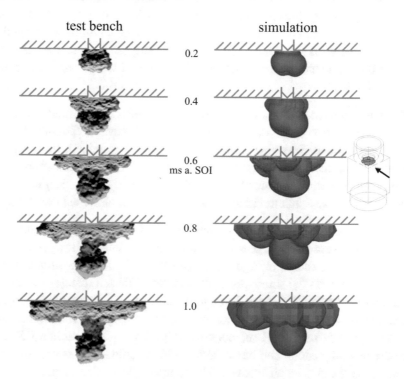

Figure 8.13: Comparison of Schlieren images to simulated spray propagation for mounting depth of -2 mm (hollowcone injector; CH4 in N2; variable spray angle α; iso-surface, threshold value = 5 % fuel concentration; l_c=1.5 mm, Δt=1/80 ms; p_{inj}=70 bar, T_f=293 K, p_{ch}=1.3 bar, T_{ch}=298 K)

[31, 46] emphasize in their work that the injector opening needs to be taken into account for a reliable injection modeling since it influences the fuel mass

flow, spray angles, velocity and droplet size for liquid fuel. The gas cloud formation can therefore also be traced back to the injector opening phase, which lasts approximately 200 to 250 μs. During this time, gaseous fuel exits the injector in a first shock wave. Due to a reduced cross section, the fuel jet can not yet reach its full spray angle. A manual determination of the angle evolution over time, using Schlieren images of the neutral injector position (0 mm), quantifies this. Initially, the spray cone angle ε is smaller than targeted, followed by continuous expansion of the spray, whereby a constant angle is reached after approximately 0.5 ms. Due to the compressibility of gaseous fuel, it requires more time to achieve a fully developed spray than the gasoline injection examined in [81].

Such an evolution was not considered initially. Instead, constant spray angles for $\alpha = 60°$ and $\gamma = 3.5°$ (derived from Schlieren images of the fully developed jet) were defined. Implementing the empirically determined, time dependent values for α, results in a significantly better conformity of the spray progression with the optical measurements as displayed in Figure 8.13. γ could not be derived from the Schlieren images and was therefore assumed constant.

Applying the temporally variable spray angles for all mounting depth simulations yields a satisfying agreement regarding the characteristic jet shapes and penetrations, as shown in Figure 8.14 and 8.15. Figure 8.16 summarizes and directly compares the simulation results for the different injector mounting positions at 0.8 ms a. SOI. In addition to the spray images presented before, also the inward shifted +2 mm position is depicted. Except for the larger distance to the upper chamber wall and consequently deeper penetration, no difference to the neutral position can be observed, which could serve as explanatory approach for the distinct difference in HC concentrations. For the outward shifted positions, in turn, it is already possible to recognize a coherent relation with the HC measurements presented above.

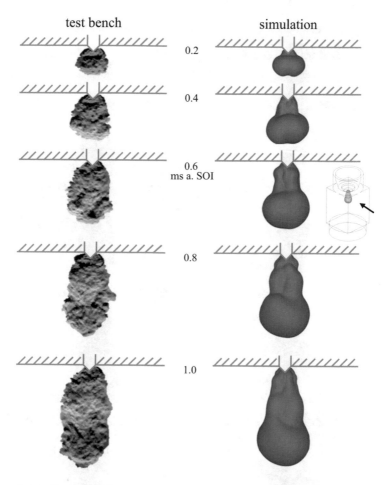

Figure 8.14: Comparison of Schlieren images to simulated spray propagation for mounting depth of ±0 mm (hollowcone injector; CH4 in N2; variable spray angle α; iso-surface, threshold value = 5 % fuel concentration; l_c=1.5 mm, Δt=1/80 ms; p_{inj}=70 bar, T_f=293 K, p_{ch}=1.3 bar, T_{ch}=298 K)

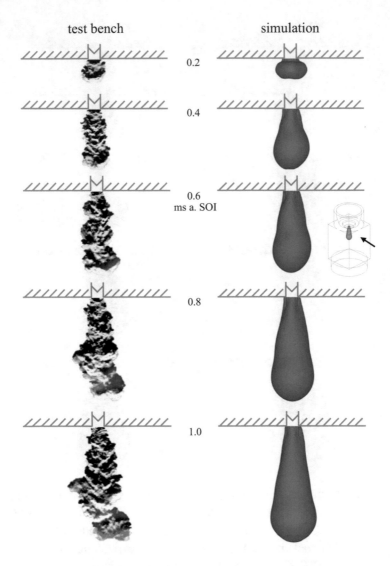

Figure 8.15: Comparison of Schlieren images to simulated spray propagation
for mounting depth of -4 mm (hollowcone injector; CH4 in N2;
variable spray angle α; iso-surface, threshold value = 5 % fuel
concentration; l_c=1.5 mm, Δt=1/80 ms; p_{inj}=70 bar, T_f=293 K,
p_{ch}=1.3 bar, T_{ch}=298 K)

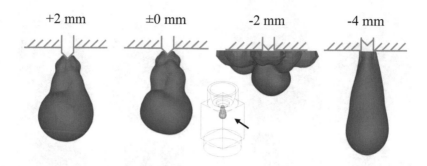

Figure 8.16: Simulation results of different injector mounting positions (hollowcone injector; CH4 in N2; variable spray angle α; iso-surface, threshold value = 5 % fuel concentration; 0.8 ms a. SOI; l_c=1.5 mm; p_{inj}=70 bar, T_f=293 K, p_{ch}=1.3 bar, T_{ch}=298 K)

Following the thorough injector calibration, the determined injection parameters were applied to a simulation of the 2-cylinder DISI Weber engine (compare to Chapter 5.4). The converged results enabled the analysis of the fuel distribution, which is illustrated in Figure 8.17. The concentration plots represent a section through the injector and one intake valve. Herewith, the fuel progression can be directly retraced. In this way the difference between the neutral and protruded injector tip becomes apparent. The latter is not influenced by the cylinder walls, collapses and penetrates into the combustion chamber. For this reason and in accordance to the test bench measurements presented in Figure 8.9, no back flow and thus HC emissions can be detected for the mounting depth of +2 mm. The neutral position (which is located closer to the oblique cylinder head) tends to partly attach to the wall, if not quite as intensive as the -2 mm mounting depth, which explicitly follows the trend observed in the injection chamber simulation. In both cases, however, fuel reaches the intake valves leading to a high fuel storage inside the channels and subsequent overflow into the exhaust system, as additionally illustrated in Figure 8.18. The strong spray collapse, caused by an outward shift of -4 mm, lets the fuel penetrate as a single beam straight towards the piston. A deflection towards the intake valves does not occur and thus no back flow of the fuel.

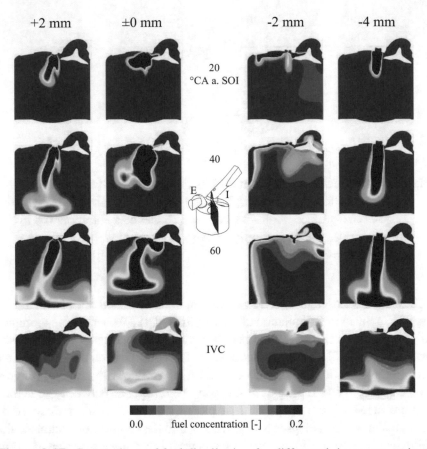

Figure 8.17: Comparison of fuel distribution for different injector mounting
positions (2-cylinder DISI Weber engine, cylinder 1; 2000 rpm,
BMEP=12 bar, p_{inj}=70 bar, T_f=293 K; SOI=220 °CA b. FTDC)

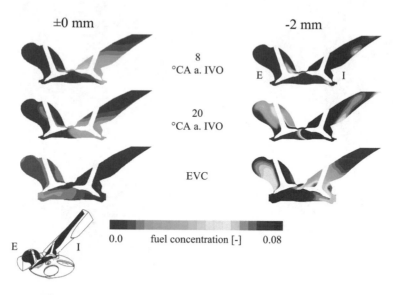

Figure 8.18: Visualization of fuel scavenging for different injector mounting positions (2-cylinder DISI Weber engine, cylinder 1; 2000 rpm, BMEP=12 bar, p_{inj}=70 bar, T_f=293 K; SOI=220 °CA b. FTDC)

Chapter 6 already highlighted the sensitivity of gaseous fuel injection towards modifications of the numerical boundary conditions. Additional injector analyses revealed that effects which again do not show remarkable influence on liquid fuel injection (at least not for high p_{inj}) must not be neglected for gaseous fuel. In contrast to the previously described deviated axis orientation, phenomena effected by the injector opening phase can theoretically be implemented in QuickSim by means of adequate models. These however require injector specific geometry parameters as input, which are generally not provided by the injector manufacturer. Thus, a thorough calibration on the basis of empirically derived parameters needs to be carried out. In order to ensure transferability, it is hereby of utmost importance that the numerical conditions of the injection chamber correspond to those of the engine simulation.

9 Conclusion and Outlook

In the present work, sensitivities and specifics of the fuel injection modeling in conformity with the 3D-CFD Tool QuickSim could be determined and best practice solutions identified. These particularly related to numerical framework conditions, the modeling of fuel and parameterization of the injector geometry.

An important finding, which can be derived from the large number of analyses, is that no generally valid procedure for the injection simulation can be defined. It is rather reasonable to identify the minimum (numerical) requirements in dependence of the specific application and focus on an adequate choice of boundary conditions. These can be determined by a comprehensive empirical calibration of the injection (preferably of macroscopic spray properties) to ensure accuracy and reliability of the simulation results. In doing so, a validation by means of reference data is mandatory. These can be optical measurements, which provide images of the fuel propagation (shadowgraphy, Mie-scattering, laser-induced fluorescence), information on the droplet sizes and velocity distribution (particle image velocimetry, phase doppler analyzer) or even the in-cylinder spray progression (transparent engine, endoscope technique). Moreover, an additional, decoupled injector internal flow simulation could provide useful information. However, this is considered to be too complex for engine development tasks and often limited in its reliability (due to a lack of geometry data or a poor choice of boundary conditions).

Following the definition of basic numerical boundary conditions (spatial and temporal discretization, droplet initialization domain, etc.), there are numerous adjusting screws to match the macroscopic spray characteristics, e.g. the number of droplets per parcels, breakup coefficients, drag coefficient, etc. In order to make the calibration process as time-efficient as possible, it would therefore be useful to develop an integrated, ideally semi-automated, calibration routine. In this way, the sequence of modifications could be designed more effectively. It would even be conceivable to implement an automated adjustment of the jet axis geometry by means of an iterative experiment-simulation adaptation.

© Springer Fachmedien Wiesbaden GmbH, part of Springer Nature 2019
M. Wentsch, *Analysis of Injection Processes in an Innovative 3D-CFD Tool for the Simulation of Internal Combustion Engines*, Wissenschaftliche Reihe Fahrzeugtechnik Universität Stuttgart, https://doi.org/10.1007/978-3-658-22167-6_9

Regarding the fuel modeling, a multi-component description of the evaporated fuel should be implemented. Up to now, this has been renounced in favor of a reduced caloric and therefore CPU-time savings. For some applications, however, a multi-component approach would be reasonable. This includes, for example, complementary water injection in SI or diesel engines for the purpose of power and efficiency increase as well as emissions reduction. This purpose requires changes in the implemented calorics in order to reliably describe the thermodynamic properties of the working fluid mixture. The same applies to simulations of dual fuel injection strategies, i.e. natural gas in combination with diesel.

Bibliography

[1] J. Abraham and V. Magi. A Model for Multicomponent Droplet Vaporization in Sprays. In *SAE Technical Paper Series*, 1998.

[2] J. D. Anderson. *Computational Fluid Dynamics*. McGraw-Hill Education - Europe, 1995.

[3] M. Baratta, A. E. Catania, E. Spessa, L. Herrmann, and K. Roessler. Multi-Dimensional Modeling of Direct Natural-Gas Injection and Mixture Formation in a Stratified-Charge SI Engine with Centrally Mounted Injector. In *SAE Technical Paper Series*, number 2008-01-0975, 2008.

[4] M. Bargende. Grundlagen der Verbrennungsmotoren. lecture notes, 2011.

[5] M. Bargende. Berechnung und Analyse innermotorischer Vorgänge. lecture notes, 2013.

[6] M. Bargende, M. Chiodi, M. Grill, and D. Wichelhaus. Virtuelle Entwicklung des Antriebsstrangs Beherrschung und Optimierung des Gesamtsystems. *ATZextra*, 2, 2016.

[7] M. Bargende, U. Riegler, and B. Scholz. Der virtuelle Motor - Fiktion oder Realität? Tagung im Haus der Technik, Essen, October 2000.

[8] J. J. Batteh and E. W. Curtis. Modeling Transient Fuel Effects with Alternative Fuels. In *SAE Technical Paper Series*, 2005.

[9] B. Befrui, G. Corbinelli, M. D'Onofrio, and D. Varble. GDI Multi-Hole Injector Internal Flow and Spray Analysis. In *SAE Technical Paper Series*, number 2011-01-1211, 2011.

[10] H.-J. Berner and M. Bargende. Erdgas als alternativer Kraftstoff - ein Überblick. Innovative Fahrzeugantriebe, Dresden, 2000.

[11] A. Böge, editor. *Handbuch Maschinenbau*. Vieweg+Teubner Verlag, 2010.

© Springer Fachmedien Wiesbaden GmbH, part of Springer Nature 2019
M. Wentsch, *Analysis of Injection Processes in an Innovative 3D-CFD Tool for the Simulation of Internal Combustion Engines*, Wissenschaftliche Reihe Fahrzeugtechnik Universität Stuttgart, https://doi.org/10.1007/978-3-658-22167-6

[12] J. Blazek. *Computational Fluid Dynamics: Principles and Applications*. Elsevier Science, 2005.

[13] G. Bruneaux, F. Grisch, S. Kaiser, and C. Schulz. Development of quantitative mixture measurement techniques for multi-component fuels including bio-fuel mmixture for automotive applications. FVV Conference on Engines, Magdeburg, Autumn 2016.

[14] Cameron Tropea et al. Short Course on Atomization and Sprays. Technische Universität Darmstadt, February 2015.

[15] CD adapco. *Methodology - STAR-CD Version 4.20*, 2013.

[16] M. Chiodi. *An Innovative 3D-CFD-Approach towards Virtual Development of Internal Combustion Engines*. PhD thesis, University of Stuttgart, 2010.

[17] M. Chiodi, A. Perrone, P. Roberti, M. Bargende, A. Ferrari, and D. Wichelhaus. 3D-CFD Virtual Engine Test Bench of a 1.6 Liter Turbo-Charged GDI-Race-Engine with Focus on Fuel Injection. *SAE International Journal of Engines*, 6(3):1834–1845, sep 2013.

[18] D. L. Christa Lüdecke. *Thermodynamik - Physikalisch-chemische Grundlagen der thermischen Verfahrenstechnik*. Springer, 2000.

[19] DDBST GmbH. Onine services. http://www.ddbst.com/online.html. Visited on 03/10/2016.

[20] T. Eder. Der Schlüssel zum Erfolg sind übergreifende, interdisziplinäre Teams. *Interview in: ATZextra – Prüfstände und Simulationen für Antriebe. In conversation with Richard Backhaus*, page 6 – 8, September 2015.

[21] FIA Sport - Technical Department. Specific Regulations for WRC. April 2014.

[22] J. Fröhlich. *Large Eddy Simulation turbulenter Strömungen*. Vieweg+Teubner Verlag, 2007.

[23] Gamma Technologies. *GT-SUITE - Flow Theory Manual*, 2016.

[24] K. Gartung. *Modellierung der Verdunstung realer Kraftstoffe zur Simulation der Gemischbildung bei Benzindirekteinspritzung.* PhD thesis, Universität Bayreuth, 2006.

[25] J. Hadler, C. Lensch-Franzen, M. Kronstedt, and M. Wittemann. Kombination von Simulation und Versuchsführung zur zielgerichteten Antriebsentwicklung. *ATZextra – Prüfstände und Simulationen für Antriebe,* page 18 – 23, September 2015.

[26] W. F. P. Helmut Eichlseder, Manfred Klüting. *Grundlagen und Technologien des Ottomotors.* Springer-Verlag KG, 2008.

[27] J. B. Heywood. *Internal Combustion Engine Fundamentals.* McGraw-Hill Education - Europe, 1988.

[28] H. L. Husted, G. Karl, S. Schilling, and C. Weber. Direct Injection of CNG for Driving Performance with low CO. 23. Aachener Kolloquium. Aachen, 2014.

[29] B. Jastrow. *Entwicklung einer Immersed Boundary Methode für einen block-strukturierten CFD-Code.* PhD thesis, Karlsruher Institut für Technologie, 2014.

[30] P. Jochmann, F. Köpple, A. Storch, A. Kufferath, B. Durst, B. Hußmann, M. Miklautschitsch, and E. Schünemann. Minimierung der Partikelemissionen von Ottomotoren mit zentraler Direkteinspritzung durch innovative Injektortechnologien und kombinierten Einsatz von CFD und Motorversuch. 10th International Symposium on Combustion Diagnostics, Baden-Baden, May 2012.

[31] D. Koch, G. Wachtmeister, M. Wentsch, M. Chiodi, M. Bargende, C. Pötsch, and D. Wichelhaus. Investigation of the Mixture Formation Process with Combined Injection Strategies in High-Performance SI-Engines. 16th Stuttgart International Symposium, Stuttgart, March 2016.

[32] T. Koch. *Numerischer Beitrag zur Charakterisierung und Vorausberechnung der Gemischbildung und Verbrennung in einem direkteingespritzten, strahlgeführten Ottomotor.* PhD thesis, Universität Karlsruhe, 2002.

[33] F. Köpple. *Untersuchung der Potentiale der numerischen Strömungs-berechnung zur Prognose der Partikelemissionen in Ottomotoren mit Direkteinspritzung.* PhD thesis, University of Stuttgart, 2015.

[34] E. Laurien and H. O. jr. *Numerische Strömungsmechanik: Grundgleichungen und Modelle - Lösungsmethoden - Qualität und Genauigkeit.* Vieweg+Teubner Verlag, 2011.

[35] C. F. Lee and F. V. Bracco. Initial Comparisons of Computed and Measured Hollow-Cone Sprays in an Engine. In *SAE Technical Paper Series*, number 940398, 1994.

[36] A. H. Lefebvre. *Atomization and Sprays.* Taylor & Francis, 1989.

[37] S. Malaguti, G. Bagli, A. Montanaro, S. Piccinini, and L. Allocca. Experimental and Numerical Characterization of Gasoline-Ethanol Blends from a GDI Multi-Hole Injector by Means of Multi-Component Approach. In *SAE Technical Paper Series*, 2013.

[38] F. Mathieu. *Laseroptische Untersuchungen des Einflusses alternativer Bio-Kraftstoffe auf die ottomotorische Gemischbildung.* PhD thesis, RWTH Aachen, 2015.

[39] F. Menzel, T. Seidel, W. Schmidt, J. Pape, and L. Stiegler. Single-cylinder engine as a tool for developing new combustion processes. *MTZ world-wide*, 67(3):6–9, mar 2006.

[40] G. P. Merker, C. Schwarz, and R. Teichmann, editors. *Combustion Engines Development - Mixture Formation, Combustion, Emissions and Simulation.* Springer-Verlag Berlin Heidelberg, 2012.

[41] G. P. Merker and R. Teichmann, editors. *Grundlagen Verbrennungs-motoren.* Springer Fachmedien Wiesbaden, 2014.

[42] National Institute of Standards and Technology (NIST) . NIST Reference Fluid Thermodynamic and Transport Properties Database (REFPROP). https://www.nist.gov/srd/refprop. Visited on 12/22/2016.

[43] National Institute of Standards and Technology (NIST). NIST research data. https://www.nist.gov/data. Visited on 12/22/2016.

[44] A. Nauwerck. *Untersuchung der Gemischbildung in Ottomotoren mit Direkteinspritzung bei strahlgeführtem Brennverfahren.* PhD thesis, Universität Karlsruhe (TH), 2006.

[45] B. Noll. *Numerische Strömungsmechanik.* Springer, 1993.

[46] B. Ofner. *Dieselmotorische Kraftstoffzerstaeubung und Gemischbildung mit Common-Rail Einspritzsystemen.* PhD thesis, Technische Universität München, 2001.

[47] M. Penzel. Genau berechnet. *Porsche Engineering Magazin 01-2015*, page 20 – 23, 2015.

[48] C. Pfeifer. *Experimentelle Untersuchungen von Einflußfaktoren auf die Selbstzündung von gasförmigen und flüssigen Brennstofffreistrahlen.* PhD thesis, Karlsruhe Institute of Technology, 2010.

[49] M. Pilch and C. Erdman. Use of breakup time data and velocity history data to predict the maximum size of stable fragments for acceleration-induced breakup of a liquid drop. *International Journal of Multiphase Flow*, 13(6):741–757, nov 1987.

[50] B. E. Poling, J. M. Prausnitz, and J. P. O'Connell. *The properties of gases and liquids.* McGraw-Hill, 5th edition, 2001.

[51] C. Pötsch, R. Kudicke, G. Wachtmeister, and D. Wichelhaus. Constraints of the combustion process for a supercharged DISI-engine for applications in motorsports. 14th Stuttgart International Symposium, Stuttgart, March 2014.

[52] Y. Ra, S. C. Kong, R. D. Reitz, C. J. Rutland, and Z. Han. Multidimensional Modeling of Transient Gas Jet Injection Using Coarse Computational Grids. In *SAE Technical Paper Series*, number 2008-01-0208, 2005.

[53] Y. Ra and R. D. Reitz. A vaporization model for discrete multicomponent fuel sprays. *International Journal of Multiphase Flow*, 35(2), feb 2009.

[54] R. D. Reitz. *Atomization and other breakup regimes in a liquid jet.* PhD thesis, Princeton University, 1978.

[55] Robert Bosch GmbH . Gasoline Systems: Piezo-Hochdruck-Einspritzventil HDEV4. http://produkte.bosch-mobility-solutions.de/media/de/ubk_europe/db_application/downloads/pdf/antrieb/de_5/gs_datenblatt_piezo_hochdruck_einspritzventil_hdev4_de.pdf. Visited on 03/10/2015.

[56] P. Roberti, M. Wentsch, M. Chiodi, and M. Bargende. Von Sieg zu Sieg mit 3D-CFD-Unterstützung. *ATZextra*, 8:48 – 53, 2015.

[57] Royal Society of Chemistry. ChemSpider - Search and share chemistry. http://www.chemspider.com. Visited on 12/16/2016.

[58] R. Scarcelli, T. Wallner, N. Matthias, V. Salazar, and S. Kaiser. Mixture Formation in Direct Injection Hydrogen Engines: CFD and Optical Analysis of Single- and Multi-Hole Nozzles. *SAE International Journal of Engines*, 4(2):2361–2375, sep 2011.

[59] T. Schlaich. Untersuchungen zum numerischen Verhalten eines Verbrennungsmodells in der 3D-CFD Simulation. Master's thesis, Universität Stuttgart, 2013.

[60] C.-O. Schmalzing. *Theoretische und experimentelle Untersuchung zum Strahlausbreitungs- und Verdampfungsverhalten aktueller Diesel-Einspritzsysteme.* PhD thesis, Universität Stuttgart, 2001.

[61] R. Schwarze. *CFD-Modellierung.* Springer Berlin Heidelberg, 2012.

[62] D. Seboldt, D. Lejsek, and M. Bargende. Untersuchungen zum Einfluss von Einblasebeginn und Einblaserichtung auf die Gemischbildung und Verbrennung an einem $\lambda = 1$ betriebenem Ottomotor mit CNG-Direkteinblasung. 10. Tagung Gasfahrzeuge, Stuttgart, 2015.

[63] D. Seboldt, D. Lejsek, and M. Bargende. Experimental Study on the Impact of the Jet Shape of an outward-opening Nozzle on Mixture Formation with CNG-DI. 16th Stuttgart International Symposium, Stuttgart, 2016.

[64] D. Seboldt, D. Lejsek, M. Wentsch, M. Chiodi, and M. Bargende. Numerical and Experimental Studies on Mixture Formation with an Outward-Opening Nozzle in a SI Engine with CNG-DI. In *SAE Technical Paper Series*, number 2016-01-0801, 2016.

[65] A.-M. Sändig. Mathematische Methoden in der Kontinuumsmechanik. lecture notes, 2005.

[66] P. Stephan, K. Schaber, K. Stephan, and F. Mayinger. *Thermodynamik - Grundlagen und technische Anwendungen Band 1: Einstoffsysteme.* Springer Berlin Heidelberg, 2013.

[67] G. Stiesch. *Elektronisches Management motorischer Fahrzeugantriebe*, chapter 4, pages 88 – 102. Vieweg+Teubner Verlag, 2010.

[68] J. Tamim and W. L. H. Hallett. Continuous thermodynamics model for multicomponent vaporization. In *Chemical Engineering Science*, volume 50, page 2933–2942, 1995.

[69] O. Tremmel. *Potenziale variabler Einspritzsysteme für die Benzin-Direkteinspritzung.* PhD thesis, Universität Hannover, 2007.

[70] R. van Basshuysen, editor. *Natural Gas and Renewable Methane for Powertrains.* Springer-Verlag GmbH, 2016.

[71] R. van Basshuysen, editor. *Ottomotor mit Direkteinspritzung und Direkteinblasung - Ottokraftstoffe, Erdgas, Methan, Wasserstoff.* Springer Fachmedien Wiesbaden, 2016.

[72] R. van Basshuysen and F. Schäfer, editors. *Handbuch Verbrennungsmotor.* Springer, 2014.

[73] VDI-Gesellschaft Verfahrenstechnik und Chemieingenieurwesen, editor. *VDI-Wärmeatlas.* Springer, 2013.

[74] J. Warnatz, U. Maas, and R. W. Dibble. *Verbrennung - Physikalisch-Chemische Grundlagen, Modellierung und Simulation, Experimente, Schadstoffentstehung.* Springer Verlag Berlin Heidelberg New York, 1997.

[75] M. Wentsch, M. Chiodi, M. Bargende, C. Pötsch, and D. Wichelhaus. Virtuelle Motorentwicklung als Erfolgsfaktor in der F.I.A. Rallye-Weltmeisterschaft (WRC). 12th International Symposium on Combustion Diagnostics, Baden-Baden, May 2016.

[76] M. Wentsch, M. Chiodi, M. Bargende, D. Seboldt, and D. Lejsek. 3D-CFD Analysis on Scavenging and Mixture Formation for CNG Direct Injection with an outward-opening Nozzle. 16th Stuttgart International Symposium, Stuttgart, March 2016.

[77] M. Wentsch, M. Chiodi, M. Bargende, D. Wichelhaus, and J. B. von Fackh. 3D-CFD-Modeling of Multi-Component Liquid Gasoline for an improved Analysis of In-Cylinder Internal Engine Phenomena. Number 216-20135172 in JSAE Annual Congress (Spring), Yokohama (Japan), May 2013.

[78] M. Wentsch, A. Perrone, M. Chiodi, M. Bargende, and D. Wichelhaus. Multi-Component Modeling of Liquid Fuel for an Improved Analysis of Fuel-Dependent Processes. 14th Stuttgart International Symposium, Stuttgart, March 2014.

[79] M. Wentsch, A. Perrone, M. Chiodi, M. Bargende, and D. Wichelhaus. Enhanced Investigations of High-Performance SI-Engines by Means of 3D-CFD Simulations. In *SAE Technical Paper Series*, number 2015-24-2469, 2015.

[80] S. Yang, Y. Ra, and R. D. Reitz. A Vaporization Model for Realistic Multi-Component Fuels. 22nd Annual Conference on Liquid Atomization and Spray Systems, Cincinnati, May 2010. ILASS Americas.

[81] C. Zuelch, A. Kulzer, M. Chiodi, and M. Bargende. The Directstart: Investigation of Mixture Formation by Means of Optical Measurements and 3D-CFD-Simulation. In *SAE Technical Paper Series*, 2005.

Appendix

Appendix 1

Table A1.1: Implemented model environment

Simulation methodology	Multiphase RANS
Turbulence	k-ε
Droplet drag	Function of the droplet Reynolds number
Secondary droplet breakup	Reitz-Diwakar
Wall impingement	Bai
Fluid	Euler-Lagrange

© Springer Fachmedien Wiesbaden GmbH, part of Springer Nature 2019
M. Wentsch, *Analysis of Injection Processes in an Innovative 3D-CFD Tool for the Simulation of Internal Combustion Engines*, Wissenschaftliche Reihe Fahrzeugtechnik Universität Stuttgart, https://doi.org/10.1007/978-3-658-22167-6

Appendix 2

Correlation equations for temperature-dependent data of liquid chemical substances [73]:

- Density ρ [kg/m^3]:

$$\rho = \frac{A}{B^{1+(1-\frac{T}{C})^D}} \qquad \text{eq. A2.1}$$

- Dynamic Viscosity η [Pa·s]:

$$\eta = e^{(A+\frac{B}{T}+C\cdot T+D\cdot T^2+E\cdot T^3)} \qquad \text{eq. A2.2}$$

- Surface Tension σ [N/m]:

$$\sigma = A\cdot(1-\tfrac{T}{T_c})^{B+C\cdot\frac{T}{T_c}+D\cdot(\frac{T}{T_c})^2+E\cdot(\frac{T}{T_c})^3} \qquad \text{eq. A2.3}$$

- Specific Heat Capacity c_p [J/kg K]:

$$c_p = A+B\cdot T+C\cdot T^2+D\cdot T^3+E\cdot T^{-2} \qquad \text{eq. A2.4}$$

- Heat of Vaporization h_v [J/kg]:

$$h_v = A\cdot(1-\tfrac{T}{T_c})^{B+C\cdot\frac{T}{T_c}+D\cdot(\frac{T}{T_c})^2+E\cdot(\frac{T}{T_c})^3} \qquad \text{eq. A2.5}$$

- Pressure of Saturation p_s [Pa]:

$$p_s = p_c\cdot e^{\frac{T_c}{T}[A\cdot(1-\frac{T}{T_c})+B\cdot(1-\frac{T}{T_c})^{1,5}+C\cdot(1-\frac{T}{T_c})^3+D\cdot(1-\frac{T}{T_c})^6]} \qquad \text{eq. A2.6}$$

Additional correlation equations for temperature-dependent data of gaseous chemical substancess [73]:

- Dynamic Viscosity η [Pa·s]:

$$\eta = A+B\cdot T+C\cdot T^2+D\cdot T^3+E\cdot T^4 \qquad \text{eq. A2.7}$$

- Specific Heat Capacity c_p [J/kg K]:

$$c_p = A+B\cdot T+C\cdot T^2+D\cdot T^3+E\cdot T^{-2} \qquad \text{eq. A2.8}$$

Appendix 3

Thermo-physical properties of single-component gasoline models at $T_f = 293$ K:

Table A3.1: Thermo-physical properties of 1c gasoline models

	1c gasoline	n-heptane	toluene	iso-pentane
ρ [kg/m^3]	738.79	686.83	869.60	623.08
η [Pa·s]	$4.38 \cdot 10^{-4}$	$4.04 \cdot 10^{-4}$	$5.85 \cdot 10^{-4}$	$2.24 \cdot 10^{-4}$
σ [N/m]	0.029	0.020	0.029	0.015
c_p [J/kg K]	2158.45	2222.04	1684.14	2272.80
h_v [J/kg]	368563.75	368563.75	415143.44	350326.25
p_s [Pa]	32576.37	4657.99	2899.10	76358.75
T_b [°C]	82	98.4	110.7	27.5

Thermo-physical properties of multi-component gasoline models at $T_f = 293$ K:

Table A3.2: Thermo-physical properties of 3c gasoline models

	n-pentane	n-octane	n-undecane
ρ [kg/m^3]	628.48	703.75	738.79
η [Pa·s]	$2.37 \cdot 10^{-4}$	$5.42 \cdot 10^{-4}$	$12.2 \cdot 10^{-4}$
σ [N/m]	0.016	0.022	0.025
c_p [J/kg K]	2290.68	2210.49	2205.01
h_v [J/kg]	371345.13	365647.04	360444.69
p_s [Pa]	56295.61	1388.67	37.03
T_b [°C]	36.1	125.7	195.9

Table A3.3: Thermo-physical properties of 5c gasoline model

	iso-pentane	cyclopentane	n-heptane
ρ [kg/m^3]	623.08	746.72	686.83
η [Pa·s]	$2.24 \cdot 10^{-4}$	$4.38 \cdot 10^{-4}$	$4.04 \cdot 10^{-4}$
σ [N/m]	0.015	0.023	0.020
c_p [J/kg K]	2272.80	1789.03	2222.04
h_v [J/kg]	350326.25	408109.41	368563.75
p_s [Pa]	76358.75	34434.65	4657.99
T_b [°C]	27.8	49.2	98.4

toluene	m-xylene
869.60	864.02
$5.85 \cdot 10^{-4}$	$6.20 \cdot 10^{-4}$
0.029	0.029
1684.14	1705.14
415143.44	404653.50
2899.10	828.41
110.7	139.1

Table A3.4: Thermo-physical properties of 7c gasoline model

	n-pentane	n-hexane	n-heptane
ρ [kg/m^3]	628.48	661.76	686.83
η [Pa·s]	$2.24 \cdot 10^{-4}$	$3.10 \cdot 10^{-4}$	$4.04 \cdot 10^{-4}$
σ [N/m]	0.016	0.018	0.020
c_p [J/kg K]	2290.68	2256.43	2222.04
h_v [J/kg]	371345.13	369445.65	368563.75
p_s [Pa]	56295.61	16199.24	4657.99
T_b [°C]	36.1	68.8	98.4

n-octane	n-nonane	n-decane	n-dodecane
703.75	718.05	729.60	699.66
$5.42 \cdot 10^{-4}$	$7.24 \cdot 10^{-4}$	$9.42 \cdot 10^{-4}$	$15.42 \cdot 10^{-4}$
0.022	0.023	0.024	0.025
2210.49	2208.14	2198.39	2199.52
365647.04	362401.52	359793.50	357241.89
1388.67	418.03	125.93	11.00
125.7	150.7	174.3	216.4